D0214387

Air Pollution, People, and Plants

An Introduction

Sagar V. Krupa
University of Minnesota
St. Paul

APS PRESS
The American Phytopathological Society
St. Paul, Minnesota, U.S.A.

Reference in this publication to a trademark, proprietary product, or company name is intended for explicit description only and does not imply approval or recommendation to the exclusion of others that may be suitable.

Library of Congress Catalog Card Number: 96-80120
International Standard Book Number: 0-89054-175-2

Printed in the United States of America on acid-free paper

The American Phytopathological Society
3340 Pilot Knob Road
St. Paul, Minnesota 55121-2097, USA

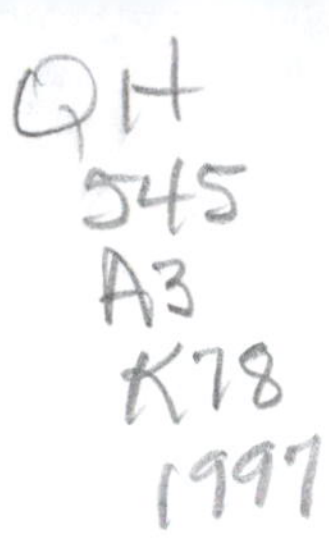

Contents

Preface

For some 25 years I have conducted scientific research on the properties of the atmosphere, the air quality, and their impacts on crops and forests. While science involves an organized expression of thinking, teaching involves an organized and clearly understandable communication of the science to others. Although in the past I have taught a graduate-level course on the subject matter of air pollution, since 1993 I have been teaching an undergraduate class on the topics discussed here, a very pleasant experience.

There are a number of books that address the broad subject of air pollution, people, and plants. However, after reviewing the contents of those books and on the basis of my own experiences, I was stimulated and challenged to prepare this introductory textbook. I have attempted to integrate the knowledge from multiple disciplines that govern the dynamic interactions between the atmosphere and the biosphere and have made every effort to describe the major aspects of the subject at a level helpful to an undergraduate student.

This effort would not have been completed without the expert help of Leslie Johnson (word processing and editorial assistance) and Sid Simms (illustrations). I am grateful to both of them. Although I do not wish to mention the names of all the individuals, at the risk of omitting someone, I thank my numerous colleagues, who over the past two decades have provided me with frequent opportunities for scientific debate and discussion, helping me to improve my knowledge of the subject matter. The products of those exchanges are reflected in this book. In addition, I am grateful to Thad Godish, who offered much of the information described in Chapters 6 and 9. I thank Robert Linderman (Senior Editor) and the editorial staff at APS Press for providing valuable support and technical help in the publication of this book. Finally, I wish to express my appreciation to the Department of Plant Pathology and the University of Minnesota Agricultural Experiment Station for extending financial assistance or support in kind in the preparation of this manuscript.

Sagar V. Krupa
February 1997

Note to the Reader

Presentation of Data on Atmospheric Trace Gas Concentrations

With the exception of atmospheric trace gas concentrations, Systèm International (SI) units have been used throughout the text to represent various quantitative measurements. For trace gas concentrations, the measurement units used are **ppm** (parts per million), **ppb** (parts per billion), and **ppt** (parts per trillion). These units represent the volume-to-volume ratio of the measured trace gas to the volume of air sampled. Such a ratio is independent of the air temperature and atmospheric pressure (elevation dependent).

Volume-to-volume ratio can also be expressed in SI units: ppm = μl per liter; ppb = nl per liter; and ppt = pl per liter. However, these units are seldom found in the published literature on air pollution effects.

Another approach is to use the SI units of the mass of trace gas (e.g., μg) to the corresponding units of air volume (e.g., m^3). This would require a systematic correction of every measurement for air temperature and atmospheric pressure (see conversion equations below). Such an effort is not possible here without introducing potentially significant mathematical errors because of the lack of original raw data for all three variables, particularly when numerous data points are used in computing average values.

Therefore, volume-to-volume ratio is used here for convenience and because it is most frequently used in the published literature from North America and for regulatory purposes by the state or provincial and federal governments there.

Conversion equations:

$$\text{ppm} = \left(\frac{\mu\text{g}}{\text{m}^3}\right)\left(\frac{0.024478}{M}\right)\left[\left(\frac{P_o}{P}\right)\left(\frac{T}{T_o}\right)\right]$$

$$\frac{\mu\text{g}}{\text{m}^3} = (\text{ppm})40.853M\left[\left(\frac{P}{P_o}\right)\left(\frac{T_o}{T}\right)\right]$$

in which

M = molecular weight of the pollutant (g/mole);
P = ambient pressure at the time of observation (atm);
P_o = pressure at standard conditions (1 atmosphere);
T = ambient temperature at the time of observation in degrees Kelvin (°C + 273.15);
T_o = temperature at standard conditions (25°C + 273.15 = 298.15°K).

Topics in Air Quality

The focus of this book is the ambient chemical climate (air pollution) and its impacts on human health and welfare (higher terrestrial plants). Because of this, three issues of air quality are not discussed here: 1) nuclear radiation; 2) noise pollution; and 3) indoor air quality. For information on these aspects of air quality, the reader should consult, among others, the following publications:

Nuclear Radiation

Cember, H. 1996. Introduction to Health Physics. McGraw-Hill, Health Professions Division, New York, NY. 733 pp.

Holmes-Siedle, A. 1993. Handbook of Radiation Effects. Oxford University Press, Oxford. 479 pp.

Noise Pollution

Saenz, A. L., and Stephens, R. W. B., eds. 1986. Noise Pollution: Effects and Control. John Wiley, Chichester, England. 446 pp.

Tempest, W., ed. 1985. The Noise Handbook. Academic Press, London. 407 pp.

Indoor Air Quality

Hines, A. L., Gosh, T. K., Loyalka, S. K., and Warder, R. C., Jr. 1993. Indoor Air: Quality and Control. PTR Prentice Hall, Englewood Cliffs, NJ. 340 pp.

Seltzer, J. M., ed. 1995. Effects of the Indoor Environment on Health. Hanley & Belfus, Philadelphia. 254 pp.

CHAPTER 1

The Atmosphere

The atmosphere is the face of the planet.

—J. E. Lovelock, *The Ages of Gaia*

Introduction

Life on earth has evolved with its climate. In this context, we frequently think of the climate in terms of its physical variables (e.g., temperature, light, and precipitation). However, climate also consists of chemical variables such as oxygen (O_2) and carbon dioxide (CO_2), gases necessary to sustain life on earth. Thus, the physical and chemical components of the atmosphere are inherently related to each other. Several gases in the atmosphere (e.g., nitrogen, N_2; oxygen; and argon, Ar) are products of evolution and have remained relatively constant in their concentrations during millions of years (National Oceanic and Atmospheric Administration, 1976). There are also a number of "trace gases" in the atmosphere with variable concentrations (Table 1.1). Although many of these trace gases are not novel to the atmosphere, significant variability in their concentrations in time and space are predominantly governed by human activity. Nevertheless, it is important to recognize that some atmospheric constituents, such as sulfur dioxide (SO_2), are produced by both natural (e.g., volcanic eruption) and human (e.g., fossil fuel combustion) activities, while others, such as chlorofluorocarbons (CFCs), are totally new and are the results of human influence. The type, nature, and sources of some trace chemical constituents in the atmosphere are discussed in greater detail in Chapter 2.

One of the definitions of **to pollute** is "to contaminate an environment, especially with human-made waste" (slightly modified from *Webster's Ninth New Collegiate Dictionary*, 1991). Therefore, we frequently think of air pollution as something strictly caused by human activity. However, in view of the earlier statement that there are both natural and human-made sources for many trace gases in the atmosphere, a more appropriate definition for an **air pollutant** should be **"a chemical constituent added to the atmosphere through human activities resulting in the elevation of its concentration above a background."** This background concentration may be zero, as in the case of CFCs, or 10 **ppb,** as in the case of ozone (O_3). Thus, the background concentrations for various air pollutants will differ depending on the nature of their origin and their amount and behavior. Nevertheless, such background values are generally derived from concentration data gathered prior to the Industrial Revolution or at geographic locations considered to be remote, even after the onset of the industrial era. The latter approach is problematic because it is highly questionable whether there are any geographic locations on the earth today where the air mass is *not* influenced by human activity. Nonetheless, in a relative sense, infor-

Table 1.1 Chemical composition of dry tropospheric air (0 to ~15 km above the earth's surface), reference year 1989[a]

Chemical constituent[b]	Formula	Mixing ratio[c]	Sources[d]
Argon	Ar	0.93%	R
Nitrogen	N_2	78.1%	B, V
Oxygen	O_2	20.9%	B
Ammonia[e]	NH_3	0.1–1 ppbv	A,B
Carbon dioxide	CO_2	354 ppmv	A, B, V
Carbon monoxide[e]	CO	40–150 ppbv[f]	A, B, P
CFC-11	$CFCl_3$	280 pptv	A
CFC-12	CF_2Cl_2	480 pptv	A
Formaldehyde	HCHO	0.1–1 ppbv	A, P
Helium	He	5.2 ppmv	R
Hydrogen	H_2	0.5 ppmv	A, B, P
Krypton	Kr	1.1 ppmv	R
Methane	CH_4	1.72 ppmv	A, B
Neon	Ne	18.2 ppmv	V
Nitric acid	HNO_3	50–1,000 pptv[f]	P
Nitric oxide[e]	NO	5–100 ppbv[f]	A, B, P
Nitrogen dioxide[e]	NO_2	10–100 ppbv[f]	B, P
Nitrous oxide	N_2O	310 ppbv	A, B, P
Ozone[e]	O_3	10–100 ppbv[f]	P
Sulfur dioxide[e]	SO_2	up to 0.2 ppbv[f]	A, P, V
Xenon	Xe	0.09 ppmv	R

[a] Source: German Bundestag, 1991.
[b] Bold lettering indicates a relatively constant mixing ratio.
[c] The ratio of the mass of a given gas to the mass of dry air. For low frequency components, mixing ratios (mole fractions) are given in ppmv (parts per million by volume) = 10^{-6}; ppbv (parts per billion by volume) = 10^{-9}; pptv (parts per trillion by volume) = 10^{-12}.
[d] A = human-made emissions; B = biosphere; P = photochemistry; R = radioactive decay; and V = volcanic activity.
[e] Trace gases with widely varying mixing ratios.
[f] Outside polluted areas where concentrations are yet higher.

mation is available on the background concentrations of many air pollutants.

Having defined air pollution and air pollutants, it is important to realize that the concern for air quality is not a new phenomenon. During the late 12th century, the philosopher, scientist, and jurist Moses Maimonides (1135–1204) wrote the following (quoted in Goodhill, 1971):

> Comparing the air of cities to the air of deserts and arid lands is like comparing waters that are befouled and turbid to waters that are fine and pure. In the city, because of the height of its buildings, the narrowness of its streets, and all that pours forth from its inhabitants and their superfluities . . . the air becomes stagnant, turbid, thick, misty and foggy. . . . If there is no choice in this matter, for we have grown up in cities and have become accustomed to them, you should . . . select from the cities one of open horizons . . . endeavor at least to dwell at the outskirts of the city.

Relationships Between Physical and Chemical Factors of the Atmosphere

Starting at the surface of the earth, the atmosphere can be divided into several distinct regions on the basis of the relationship between temperature and altitude (Fig. 1.1). The temperature initially falls with increasing altitude up to about 10–12 km (the **troposphere**). This is the reason that as you climb increasingly higher up the mountains, it becomes progressively cooler. At ~15 km above the earth's surface, this trend is reversed at the transition zone between the troposphere and the stratosphere known as the **tropopause**. In the **stratosphere** (~15–50 km above the surface), the temperature increases with altitude. From 50 to ~85 km in the region known as the **mesosphere**, the temperature drops again and then begins to rise once more in the **thermosphere** (>85 km).

The drop in temperature with increasing altitude in the troposphere (~0–15 km) results from the strong heating effect at the surface of the earth from the absorption of visible and near-ultraviolet (UV) solar radiation (see Table 1.2 for the properties of incoming solar radiation). Accompanying this is a strong vertical mixing in the atmosphere that causes particulate and gaseous pollutants to rise from the surface to the top of the troposphere in a few days or less, depending on the meteorological conditions. Almost all of the water vapor, clouds, and precipitation (removal process of pollutants to the earth) in the atmosphere are found in this region (Finlayson-Pitts and Pitts, 1986).

In the stratosphere (~15–50 km above the surface), a series of **photochemical** (light-driven) reactions involving O_3 **(ozone)** and molecular O_2 occur. Ozone strongly absorbs solar radiation in the region of ~210–290 nm, whereas O_2 absorbs at ≤200 nm. The absorption of light, primarily by O_3, is a major factor in the increase in temperature with altitude in the stratosphere (Finlayson-Pitts and Pitts, 1986). Excited O_2 and O_3 molecules photodissociate (dissociate in the presence of light), initiating a series of reactions in which O_3 is both created and destroyed, leading to its **steady state** concentration (Notes 1.1). This is essential for life on earth, because O_3 strongly absorbs light at ≤290 nm, preventing the incidence of biologically harmful solar radiation at the surface (refer to Chapter 5 on global climate change for more details on the observed and/or predicted changes in the beneficial O_3 layer and their consequences).

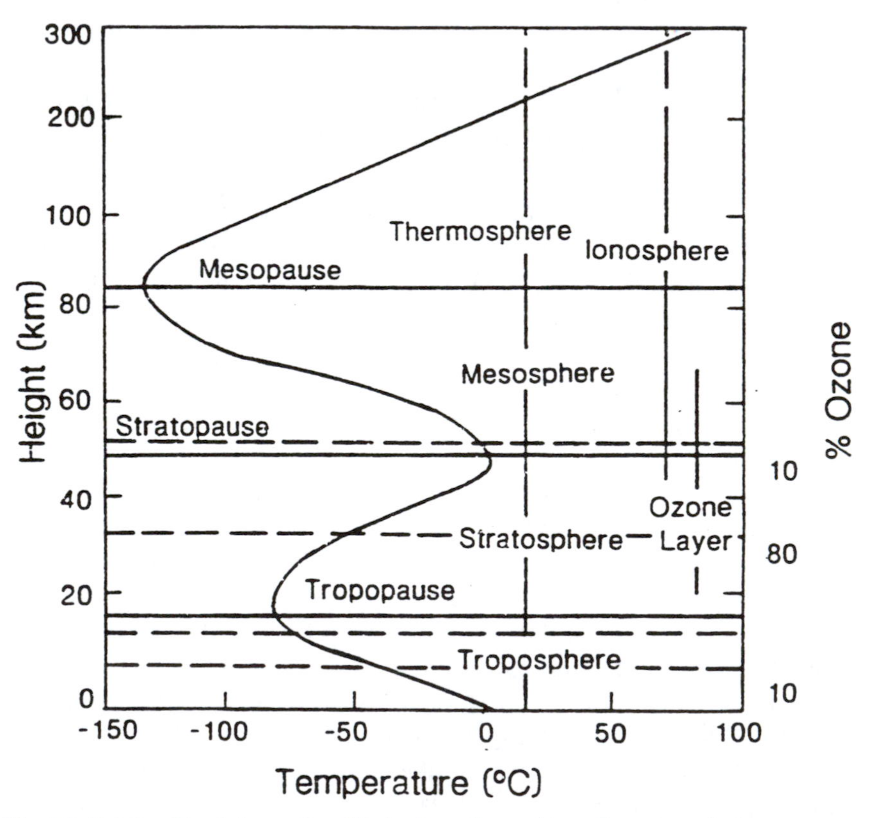

Fig. 1.1. Relationships between the altitude above the earth's surface, atmospheric temperature, and percentage of O_3 by volume.

Relatively little vertical mixing occurs in the stratosphere, and no removal of pollutants by precipitation occurs in this region. As a result, massive injections of particles, for example, from volcanic eruptions (such as that of Mount Piñatubo), often produce layers of particles in the stratosphere that persist for long periods of time, even from one to several years (Finlayson-Pitts and Pitts, 1986).

In the mesosphere (~50–85 km above the earth's surface), the temperature falls again with altitude, and vertical mixing also occurs within this region. This temperature trend is caused by the reduction in the O_3 concentration with altitude in this layer of the atmosphere, which in turn reduces the heat released through reactions 1 through 4 shown in Notes 1.1.

Above ~85 km, the temperature rises again because of the increased absorption of solar radiation at wavelengths ≤200 nm by O_2 and N_2, as well as by their atomic species, O and N. This region is known as the thermosphere.

The transition zones between various regions of the atmosphere are known as the **tropopause**, **stratopause**, and **mesopause** (Fig. 1.1). Their locations are not fixed and vary with latitude, season, and year (Finlayson-Pitts and Pitts, 1986).

Table 1.2. Breakdown into wavelength regions of relative solar radiation arriving at the earth's surface[a]

Wavelength (nm)[b]	Description	Percentage
<280	UV-C	0
280–320	UV-B	0.5
320–400	UV-A	5.6
400–800	Visible light	51.18
>800	Infrared radiation	42.1

[a] Source: German Bundestag, 1991.
[b] 1 nm = 10^{-9} meter.

Notes 1.1. Ozone formation in the stratosphere

$$O_2 + h\nu \rightarrow 2O \quad (1)$$
$$O + O_2 \xrightarrow{M} O_3 \quad (2)$$
$$O + O_3 \rightarrow 2O_2 \quad (3)$$
$$O_3 + h\nu \rightarrow O_2 + O \quad (4)$$

$h\nu$ = light energy; M = matter, a third body; O = singlet or atomic oxygen; O_2 = molecular oxygen; O_3 = ozone.

While the preceding narrative provides a basic description of some of the relationships between the physical and chemical climatology in various layers of the atmosphere, it is the tropospheric processes that are immediately relevant to life on earth. Numerous air pollutants are directly emitted into the atmosphere through natural and human-related activities (Finlayson-Pitts and Pitts, 1986). These pollutants are called **primary pollutants** and exist either as gases or as liquid or solid particles. In this context, the term **aerosol** is frequently used to define liquid or solid particles in a gaseous (air) atmosphere. Theoretically, a football thrown by a quarterback during a game of American football can be classified as an aerosol until it is caught by a receiver (same or opposite team) or until it lands on the ground. Similarly, a raindrop is an aerosol until it falls onto a surface. In reality, however, the properties of aerosols are much more complex and are discussed in greater detail in later chapters.

Once a primary pollutant is emitted into the atmosphere, it is deposited onto surfaces by **dry deposition** during periods of no precipitation and by **wet deposition** during periods of precipitation (e.g., rain, fog, and hail). Dry deposition involves diffusion (gases), Brownian motion (**fine particles**, <2.5 μm in diameter), gravitational settling or sedimentation, and impaction (**coarse particles**, >2.5 μm in diameter). Wet deposition involves **in-cloud (rainout)** and **below-cloud (washout)** removal of chemical constituents (Finlayson-Pitts and Pitts, 1986).

While deposition processes govern the lifetime of pollutants in the atmosphere, depending on the meteorological conditions (e.g., vertical mixing and wind direction and speed), many air pollutants can be transported from a few to several thousand kilometers. Similarly, depending on the meteorological conditions and the physical and chemical characteristics, primary pollutants such as nitrogen dioxide (NO_2) and sulfur dioxide (SO_2) can be converted in the atmosphere into **secondary pollutants** such as ozone (O_3) and sulfate (SO_4) particles, respectively. These specific reactions are largely dependent on radiation and temperature (Finlayson-Pitts and Pitts, 1986; National Research Council, 1991).

Air pollutants are deposited onto surfaces, and they can also be transported upward in the atmosphere. Particulate matter in the atmosphere serves as **condensation nuclei** for moisture and thus cloud formation. Some forms of particulate matter are highly hygroscopic (e.g., sulfuric acid, $\mathbf{H_2SO_4}$, aerosol) and contribute to the "acid rain" or "acidic precipitation" phenomenon. The relationship between particulate matter in the atmosphere and cloud formation serves as the basis for artificial

cloud seeding (e.g., release of AgI, silver iodide particles) to generate rain. Furthermore, hygroscopic aerosols in the atmosphere are highly refractive of light. Therefore, during the summer, an accumulation of aerosols and water vapor in the atmosphere leads to reductions in visibility (Finlayson-Pitts and Pitts, 1986). Thus, visitors to the Grand Canyon, Arizona, or the Pacific Northwest (e.g., Oregon or Washington) during the summer may be subjected not only to atmospheric haze but also to optical refraction of incident light, giving false or displaced locations of distant landmarks.

Stratospheric O_3 acts like a shield blocking biologically harmful solar UV radiation and also initiates key stratospheric chemical reactions. It transforms solar radiation into heat, and the subsequent differences in the air pressure induce the mechanical energy of atmospheric winds. Thus, large-scale horizontal air movements result from the alternating heating and cooling of the earth's surface and the consequent differences in atmospheric pressure. Given the Coriolis force (the influence of the angle of the earth's rotation), jet streams move toward the pole from west to east north of the equator. The converse (east to west) is true for regions south of the equator. These phenomena represent the **general circulation patterns** across the earth's surface that subsequently result in a **polar vortex**. Certain O_3-destroying chemical constituents drift toward the pole and form one of the physical mechanisms for the observed decreases in the beneficial O_3 column, the **ozone hole** at the South Pole and possibly at the North Pole (this subject is discussed in greater detail in Chapter 5, Air Pollution and Global Climate Change).

Closer to the surface, **high** and **low pressure systems** (Notes 1.2) and **isobars** (geographic regions of equal atmospheric pressure, either high or low) lead to curved airflows. The winds associated with isobars are called **gradient winds**. Airflow in the northern hemisphere is counterclockwise (cyclonic) in a low pressure center and clockwise (anticyclonic) in a high pressure center (Anthes et al., 1981). According to the model most commonly accepted by operational meteorologists, a **cold front** is the line of intersection on the surface at which cold air from the west overtakes and flows beneath the warm air to the east in the United States.

Similarly, a **warm front** is the line at which warm air overtakes and flows over cooler air to the east. When the two fronts merge, an **occluded front** is formed. **Stationary fronts** are composed of a combination of regional-scale cold and warm air masses (Notes 1.2).

During the summer, stationary fronts result in **stagnant air masses** and **episodes** of secondary air pollutants such as $\mathbf{O_3}$ and **sulfate** $\mathbf{(SO_4)}$

particles. Such episodes can last from one to several days or longer. Below 1 km altitude, **friction** can retard the airflow and direct it toward the area of low pressure. Friction has its greatest impact near the ground (because of factors such as surface roughness and the nature of the terrain) and decreases with altitude. Over a rough terrain, wind speeds can be reduced by as much as 50% and may be inclined from the isobars by 30–50°. These phenomena further enhance the influence of stagnant air masses.

In contrast to the general circulation and regional-scale meteorology, at a much more local scale **atmospheric inversions** regulate high concentrations or episodes of pollutants (e.g., SO_2 from a coal-fired power plant) at the surface. Normally, temperature **decreases** at a rate of 1°C per every 100 meters or 5.4°F per every 1,000 feet **increase** in altitude from the surface (**normal adiabatic lapse rate**). However, temperature can also **increase** with altitude (**inversion**). Such inversions

Notes 1.2. Meteorological processes at the surface

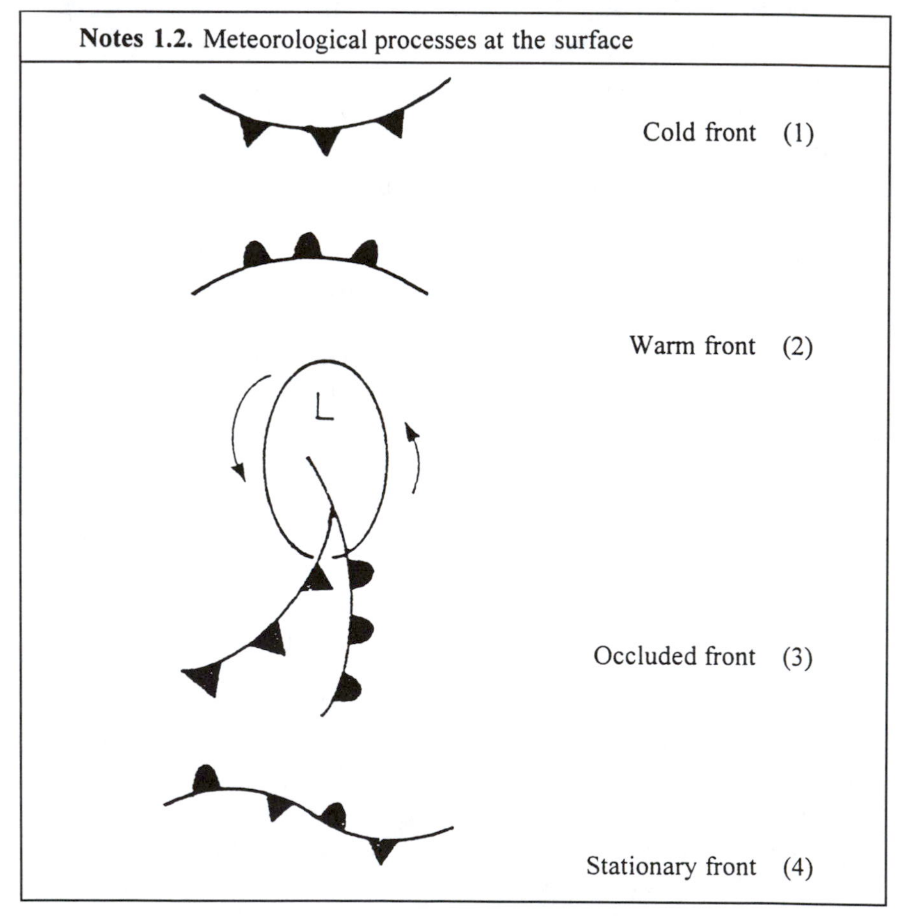

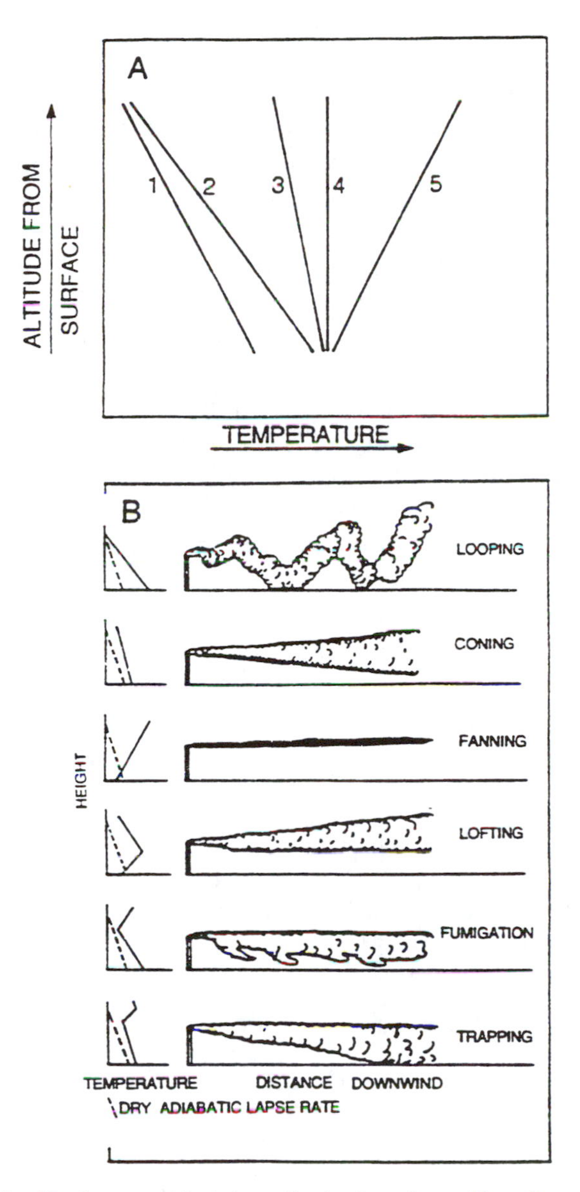

Fig. 1.2. A, Relationships between altitude immediately above the earth's surface and variations in temperature. 1 = **Normal adiabatic lapse rate** (temperature decreases at a rate of 1°C/100 m or 5.4°F/1,000 ft). 2 = **Super-adiabatic lapse rate** (temperature decreases at a rate >1°C/100 m). 3 = **Sub-adiabatic lapse rate** (temperature decreases at a rate <1°C/100 m). 4 = **Isothermal** (constant temperature with change in height). 5 = **Surface-based inversion** (temperature increases with height). **B,** Relationships between temperature profiles at different heights and the type of plume dispersion from a stationary point source. Dashed line = the standard reference of normal adiabatic lapse rate (1°C/100 m), and the solid line = the actual measured temperature profile. Conditions favoring **looping**, **fumigation,** and **trapping** will force high concentrations of pollutants to ground level.

are governed by the surface and air temperatures and can occur and vary within various layers of air close to the ground (Fig. 1.2). Figure 1.2A shows the basic relationships between temperature and altitude from the surface, and Figure 1.2B shows various plume dispersion patterns from a stationary point source (e.g., a coal-fired power plant). The type of plume dispersion pattern is dependent upon the type of temperature profile below and above the source (Fig. 1.2B, left). For example, looping occurs during conditions of super-adiabatic lapse rate; similarly, fumigation occurs during periods of super-adiabatic lapse rate at the surface followed by inversion in the air layer above the source. It is important to note that these temperature profiles are highly variable during a given day, changing after a number of minutes to hours. Nevertheless, as shown in Figure 1.2B, atmospheric conditions favoring **looping**, **fumigation,** and **trapping** of a plume will support high primary pollutant concentrations **(episodes)** at the ground level.

In contrast to surface-based inversions, **subsidence** or inversions aloft (exhibiting a sharp increase in temperature and decrease in relative humidity) trap air masses over large areas and result in regional-scale pollution (e.g., ozone, a secondary air pollutant formed in the atmosphere and a problem in Los Angeles and the entire South Coast Air Basin of California).

References

Anthes, R. A., Cahir, J. J., Fraser, A. B., and Panofsky, H. A. 1981. The Atmosphere. 3rd ed. Charles E. Merrill, Columbus, OH.

Finlayson-Pitts, B. J., and Pitts, J. N., Jr. 1986. Atmospheric Chemistry: Fundamentals and Experimental Techniques. John Wiley & Sons, New York.

German Bundestag. 1991. Protecting the Earth—A Status Report with Recommendations for a New Energy Policy. Deutscher Bundestag, Referat Öffentlichkeitsarbeit, Bonn, Germany.

Goodhill, V. 1971. Maimonides—Modern Medical Relevance. XXVI Wherry Memorial Lecture. Trans. Am. Acad. Ophthalmol. Otolaryngol. 75:463-491.

Lovelock, J. E. 1989. The Ages of Gaia. A Biography of Our Living Earth. Oxford University Press, Oxford.

National Oceanic and Atmospheric Administration. 1976. U.S. Standard Atmosphere. NOAA S/T 76-1562.

National Research Council. 1991. Rethinking the Ozone Problem in Urban and Regional Air Pollution. National Academy Press, Washington, DC.

Further Reading

McElroy, M. B. 1994. Climate of the earth: An overview. Environ. Pollut 83:3-22.

Meszaros, E. 1993. Global and Regional Changes in Atmospheric Composition. Lewis Publishers, Boca Raton, FL.

CHAPTER 2

Nature, Types, Sources, and Ambient Concentrations of Air Pollutants

> But, without the use of calculations it is evident to every one who looks at the yearly Bill of Mortality, that near half the children that are born and bred in London die under two years of age. Some have attributed this amazing destruction to luxury and the abuse of spirituous liquors: These, no doubt are powerful assistants; but the constant and unremitting poison is communicated by the foul air, which, as the town still grows larger, has made regular and steady advances in its fatal influence.
>
> —John Evelyn, *Fumifugium;* quoted by J. P. Lodge, Jr.

Introduction

As stated in Chapter 1, air pollutants occur either as gases or as particulate matter (both solid and liquid states), and they are categorized as either **primary** pollutants (emitted directly into the atmosphere by a source, e.g., sulfur dioxide [SO_2] and nitrogen dioxide [NO_2]) or **secondary** pollutants (formed secondarily in the atmosphere through reactions involving primary emissions, e.g., ozone [O_3] and sulfate [SO_4] particles).

Table 2.1. Air pollutants and related processes as issues in health and welfare

Restricted local scale	Local scale	Regional-continental scale	Global scale
Ammonia	Coarse particulate aerosols	Acidic precipitation	Carbon dioxide
Chlorine, hydrogen chloride	Hydrogen fluoride	Fine particulate aerosols [a]	Fine particulate aerosols [a]
Ethylene	Sulfur dioxide	Ozone	Moisture
Nitrogen dioxide		Peroxyacetyl nitrate (possibly)	Temperature
			UV-B radiation [b]
			Winds

[a] Includes sulfate, nitrate, ammonium, organics, and trace metals.
[b] 280–320 nm (1 nm = 10^{-9} meter).

Depending on their physical and chemical properties and on the prevailing local, regional, or even interregional meteorological conditions, primary pollutants can be deposited onto surfaces in proximity to their sources or transported from a few to thousands of kilometers away. During long-range transport, primary pollutants can be transformed in the atmosphere into secondary pollutants. Therefore, air pollutants are classified into categories based on their geographic scale of environmental importance (Table 2.1).

Table 2.2. Selected air pollutants and their sources of origin through human activity

Air pollutant	Some sources of origin
Ammonia (NH_3)	Leaks or breakdown in industrial operations; spillage of anhydrous ammonia; feedlots and stockyards
Carbon dioxide (CO_2)	Fossil fuel combustion; land-use conversion
Carbon monoxide (CO)	Transportation; fossil fuel combustion; solid waste disposal; agricultural burning; steel production; photolysis of methane
Chlorine (Cl) and hydrogen chloride (HCl)	Petroleum refineries; glass industries; plastic incineration; scrap burning; accidental spills
Ethylene (C_2H_4) and related olefins[a] or hydrocarbons (HCs)	Motor vehicles; refuse burning; combustion of coal and oil; leaky natural gas heaters
Hydrogen fluoride (HF)	Aluminum industries; steel manufacturing industries; phosphate fertilizer plants; brick plants
Lead (Pb)	Oil and gasoline combustion; coal combustion; waste incineration; wood combustion; iron and steel production; secondary nonferrous metal production; primary copper smelting; mining of nonferrous metals; primary lead smelting; primary zinc smelting; primary nickel smelting
Metals (Me) (other than lead)	Same as sources of particulate matter (with the exception of reactions in the atmosphere); industry involving some form of combustion (e.g., power plants and metal smelters)
Organic compounds (OCs)	Pesticides; solvents; electric transformer insulators; fire retardants; refrigerants; degreasing agents; foaming agents; aerosol propellants; wood burning
Oxides of nitrogen (NO_x) (nitrogen dioxide, NO_2 + nitric oxide, NO)	Automobiles and trucks; combustion of natural gas, fuel oil, and coal; refining of petroleum; compressors of natural gas; incineration of organic waste; wood burning
Ozone (O_3)[b]	Photochemical reactions in the atmosphere from primary precursor pollutants
Particulate matter (PM)	
Fine (<2.5 μm)[b]	Both a primary and a secondary pollutant. Predominant portion from homogeneous and heterogeneous reactions in the atmosphere; coal and fossil fuel combustion; metal smelting and refining processes; a variety of other industries including waste incineration; wood burning
Coarse (>2.5 μm)	Cement mills; lime kilns; incinerators; combustion of coal, gasoline, and fuel oil; a variety of industries; wood burning; agricultural practices
Peroxyacetyl nitrate (PAN)[b]	Photochemical reactions in the atmosphere
Sulfur dioxide (SO_2)	Combustion of fossil fuels; petroleum and natural gas industries; metal smelting and refining processes

[a] **Olefin** = straight-chained or branched hydrocarbon containing one or more double bonds.
[b] Secondary pollutant. **Heterogeneous reaction** = gas-to-particle phase reaction; **homogeneous reaction** = gas-to-gas phase reaction; **photochemical reaction** = chemical reaction driven by sunlight.

The sources of these air pollutants can be classified simply as **stationary** or **mobile** by their nature. Stationary sources are further classified as **single-event point** (an accidental chemical spill at one time at a given geographic location, e.g., derailment of a chemical container); **continuous point** (a chimney at a given geographic location that emits air pollutants continuously, e.g., a coal-fired power plant); **area** (a large urban center, e.g., Minneapolis-St. Paul, Minnesota); **regional** (several closely located urban centers, e.g., Los Angeles and the entire South Coast Air Basin in California); or **continental** (many adjacent and closely located nations, each with multiple urban centers, e.g., Europe). Similarly, mobile sources are viewed as **line** sources (pollutant emission that occurs in a line, e.g., a highway with automobiles traveling in a line or an aircraft jet stream) (Skelly et al., 1979).

Sources of Primary Pollutants in the Atmosphere

As stated in Chapter 1, many primary pollutants are emitted by both natural and human-made sources. The strength or importance of the two types of sources varies with the pollutant under consideration. For example, while fossil fuel combustion is a major source of sulfur dioxide in the atmosphere, geo-biogenic (natural) sources are major contributors of atmospheric methane, CH_4, a chemical constituent of much concern in global warming.

Table 2.2 provides a summary of selected air pollutants and their major sources of origin through human (**anthropogenic**) activity. An

Table 2.3. Estimates of global emission to the atmosphere of gaseous sulfur compounds [a,b]

Source[c]	**Annual flux (TgS)**[d]
Anthropogenic (mainly SO_2 from fossil fuel combustion)	80
Biomass burning (SO_2)	7
Oceans (DMS)	40
Soils and plants (H_2S, DMS)	10
Volcanoes (H_2S, SO_2)	10
Total	147

[a] Source: Houghton et al., 1990.
[b] The uncertainty ranges are estimated to be about 30% for the anthropogenic flux and a factor of two for the natural fluxes.
[c] DMS = dimethyl sulfide; H_2S = hydrogen sulfide; and SO_2 = sulfur dioxide.
[d] Teragram (Tg) sulfur (S) = 10^{12} or 1 trillion grams of sulfur.

analysis of the information in the table shows that transportation, fossil fuel combustion for uses other than transportation, and the use of chemicals in agriculture and in numerous types of industries are the common sources of many primary pollutants. Tables 2.3 and 2.4 provide summary statistics of global emissions of two key primary pollutants, sulfur dioxide (SO_2) and the oxides of nitrogen (NO_x, $NO + NO_2$). It can be concluded from Table 2.3 that both natural (geo-biogenic) and human-made (anthropogenic) sources are important in the global-scale atmospheric sulfur burden (60 and 87 Tg [teragrams], respectively, per year). Similarly, the data presented in Table 2.4 show that human activity (fossil fuel combustion, including transportation and biomass burning) contributes significant amounts of nitrogen to the atmosphere, in fact as much as natural processes (lightning, soil microbial processes, and the oxidation of ammonia). In comparison, Table 2.5 shows that natural processes (e.g., wetlands and rice paddies) are more important

Table 2.4. Global budget for the oxides of nitrogen, NO_x ($NO + NO_2$)[a,b]

Source category	Nitrogen[c] (10^{12} g year^{-1})
Fossil fuel combustion	21 (14–28)
Biomass burning	12 (4–24)
Lightning	8 (2–20)
Microbial activity in soils	8 (4–16)
Oxidation of ammonia	1–10
Photolytic[d] or biological processes in the ocean	>1
Input from the stratosphere	~0.5
Total	25–99

[a] Adapted from Logan, 1983.
[b] NO = nitric oxide; and NO_2 = nitrogen dioxide.
[c] Ranges are given in parentheses.
[d] Decomposition in the presence of light.

Table 2.5. Estimated sources of methane (CH_4)[a]

Source	Annual release ($TgCH_4$)[b]
Natural wetlands (bogs, swamps, tundra, etc.)	115
Rice paddies	110
Enteric fermentation (animals)	80
Gas drilling, venting, transmission	45
Biomass burning	40
Termites	40
Landfills	40
Coal mining	35
Oceans	10
Fresh waters	5
Total	520

[a] Adapted from Houghton et al., 1990.
[b] Teragram (Tg) methane (CH_4) = 10^{12} or 1 trillion grams of methane.

than direct anthropogenic sources in the emission of methane, CH_4, a greenhouse or global warming gas. Thus, as previously stated, natural or human-made sources can vary in their importance to the contribution of the atmospheric burden of a chemical constituent or to air quality, depending on the pollutant in question.

An interesting point can be made regarding another important air pollutant, carbon monoxide (CO). We frequently link CO to automobile exhaust and to effects on human health. This is one of the reasons for automobile emission testing in the United States. Automobiles tend to emit more CO during extremely cold winter months because of hard starts and stops. Therefore, we are required to use oxygenated fuels (gasoline containing ethanol, for example) during those months to achieve complete fuel combustion in order to reduce CO emissions. Even though the automobile is a major source of CO, there are several natural sources for the pollutant (Table 2.6). As opposed to the CO from the automobile exhaust, which remains at the surface where we breath the air, many natural sources contribute to the CO burden of the entire troposphere.

While the preceding narrative emphasizes some of the gaseous primary pollutants in the atmosphere, particulate matter in the ambient air can also be both primary or secondary in its origin and can be in either a dry or a wet state. For convenience and continuity, certain characteristics of both **primary** and **secondary particulate matter** are described in the following section of this chapter and other aspects are discussed in Chapter 4, Transport of Air Pollution Across Regional, National, and International Boundaries.

Table 2.6. Estimated range of global emissions of carbon monoxide (CO) from natural and anthropogenic sources[a]

Source category[b]	Estimated emissions (10^6 metric tons)
Natural	
CH_4 oxidation	60–5,000
Oxidation of natural hydrocarbons	50–1,300
Microbial activity in oceans	20–200
Emissions from plants	20–200
Total	150–6,700
Anthropogenic	
Fossil fuel combustion, including gasoline	250–1,000
Forest fires	10–60
Total	260–1,060

[a] Source: Informatics, 1979.
[b] CH_4 = methane.

Sources of Secondary Pollutants in the Atmosphere

Gases

Secondary pollutants are produced in the atmosphere through reactions that are described as **homogeneous** (gas-to-gas phase) and **heterogeneous** (gas-to-particle phase, particles both dry and wet, e.g., cloud water or fog droplets). The most important mechanism driving homogeneous reactions at the surface is solar radiation (Notes 2.1). Because of this, secondary pollutant concentrations in the air will be high during the summer months (Fig. 2.1) when radiation and temperature are high (Finlayson-Pitts and Pitts, 1986). In this context, perhaps the most important and pervasive type of secondary gaseous air pollutant at the surface is ozone (O_3). As previously noted, this O_3 is mainly produced by human activity (not to be confused with the naturally produced, beneficial O_3 layer in the stratosphere).

The keys to elevated surface-level O_3 are chemical reactions that convert NO to NO_2 without consuming O_3 (Notes 2.1, reaction 3). Such shifts in O_3 chemistry occur in the presence of peroxy radicals (RO_2) produced by the oxidation of hydrocarbons.

As shown in Table 2.2, fossil fuel combustion (including transportation) is a major and common source for both oxides of nitrogen, NO_x (NO + NO_2) and hydrocarbons (HCs). As a matter of fact, downwind from many urban centers in the world, O_3 concentrations frequently reach high levels (Notes 2.2) during the summer months in the suburban and rural areas (National Research Council, 1991). In many urban centers across the world, peak highway traffic levels are recorded

Notes 2.1. Ozone formation in the troposphere and its consumption

$$NO_2 + h\nu \rightarrow NO + O \quad (1)$$
$$O + O_2 + M \rightarrow O_3 + M \quad (2)$$
$$O_3 + NO \rightarrow NO_2 + O_2 \quad (3)$$

Dynamic equilibrium

$$NO_2 + O_2 \overset{h\nu}{\rightleftharpoons} NO + O_3 \quad (4)$$

$h\nu$ = light energy; M = matter, a third body; NO = nitric oxide; NO_2 = nitrogen dioxide; O = singlet or atomic oxygen; O_2 = molecular oxygen; O_3 = ozone.

between 7:00 and 9:00 A.M. The emissions from this traffic and the resulting atmospheric accumulation of NO_x and HCs occur during a period of relatively low solar radiation and temperature. Minimal wind speed of, for example, 8 km per hour can push or transport this pollutant-laden air mass at least 40 km downwind by 2:00 P.M., when the radiation and temperature start to increase significantly. Thus, during summer months, high O_3 concentrations occur between 2:00 P.M. and 6:00–7:00

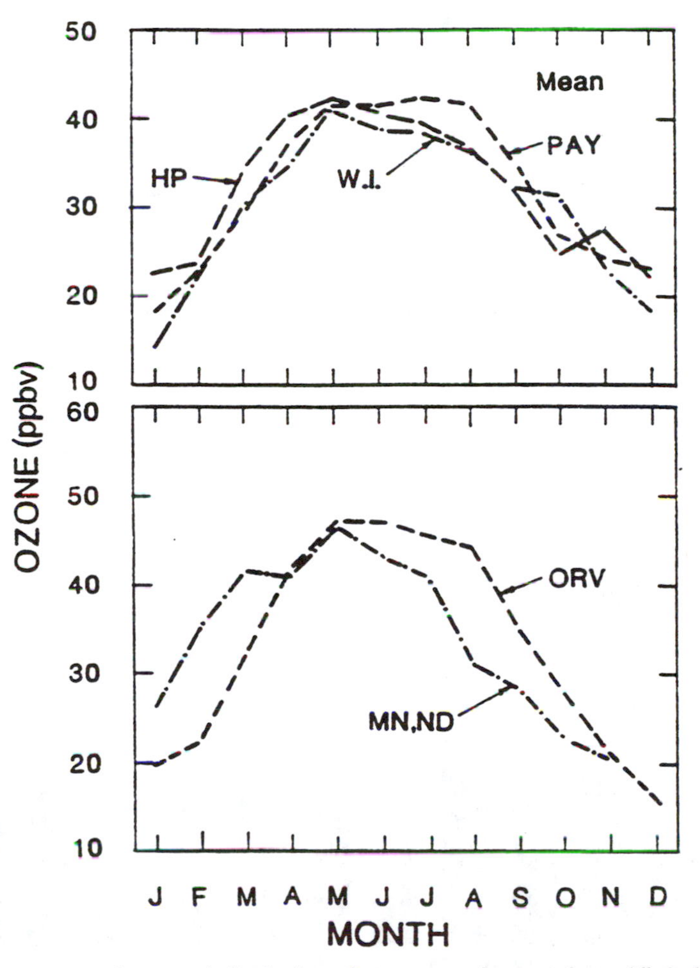

Fig. 2.1. Upper panel, seasonal distribution of ozone near the ground at midlatitudes. **Daily average** values were derived for Hohenpeissenberg **(HP),** Germany, and for Payerne **(PAY),** Switzerland, by using an **average of data from ~0930** and **~1600 hours**. Daily average values for Wallops Island **(W.I.)** in the Atlantic Ocean off the coast of Virginia were derived by using the diurnal variation of rural ozone in the eastern United States. Lower panel, results from continuous surface measurements at rural sites in the United States. Daily average values are shown for two sites in Minnesota **(MN)** and one in North Dakota **(ND)** for 1977–1981 and three sites in the Ohio River Valley **(ORV)** for May 1980 to August 1981. (Reprinted, by permision, from Logan, 1985)

P.M. in rural areas as much as 65 km downwind from, for example, Minneapolis-St. Paul, Minnesota. Maimonides (see Chapter 1), of course, was not faced with this problem and therefore suggested ". . . endeavor at least to dwell at the outskirts of the city." Present-day humans, however, living in the countryside and assuming they are breathing fresh country air may actually be subjected to more O_3 pollution than the

Notes 2.2. Ozone formation in the troposphere without its consumption	
$RO_2 + NO \rightarrow NO_2 + RO$	(1)
$NO_2 + h\nu \rightarrow NO + O$	(2)
$\mathbf{O + O_2 + M \rightarrow O_3 + M}$	(3)
$RO_2 + O_2 + h\nu \rightarrow RO + O_3$	(4)
$h\nu$ = light energy; M = matter, a third body; NO = nitric oxide; NO_2 = nitrogen dioxide; O = singlet or atomic oxygen; O_2 = molecular oxygen; O_3 = ozone; RO = oxyradical; RO_2 = peroxy radical.	

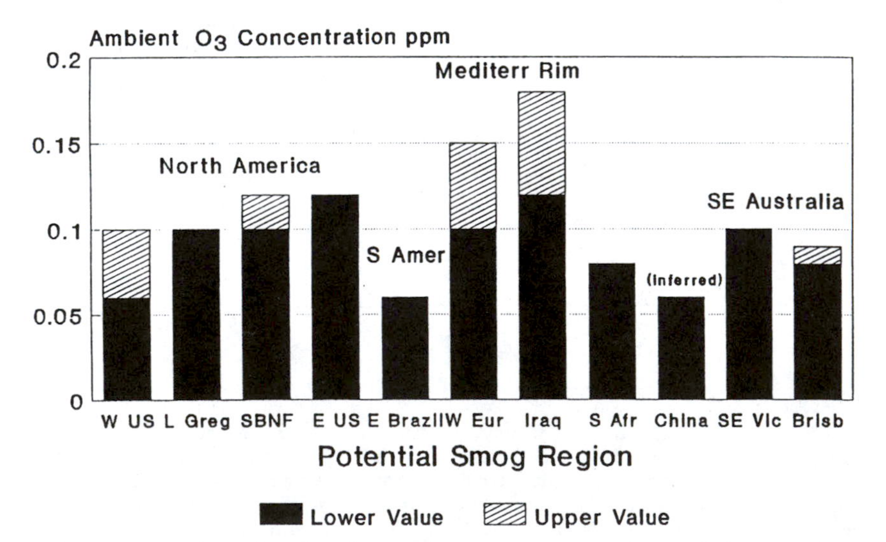

Fig. 2.2. Concentrations of surface O_3 at various locations across the world. Lower and upper values represent the range of average hourly peak O_3 concentrations exceeded more than once during the summer.

people residing in the city itself. It is extremely important, however, to note that the people dwelling in the city may be subjected to high concentrations of primary air pollutants (e.g., trace metals).

Oxides of nitrogen (NO_x) and HC concentrations also accumulate during evening traffic hours (e.g., 5:00–7:00 P.M). This air mass is pushed downwind during nighttime hours, and the same process previously described for the daylight hours can occur the next day in rural areas. Thus, surface-level O_3 is a regional- or even continental-scale problem (Fig. 2.2).

Surface-level ozone is a major component of the **photochemical oxidant complex** (an oxidant may be defined as a substance capable of oxidizing another, which is present in a reduced form) or of **photochemical smog** (a combination of smoke and fog, mainly produced through photochemical reactions). In addition to O_3, a secondary pollutant in photochemical smog is **peroxyacetyl nitrate** (PAN) (Notes 2.3)

In general, PAN occurs at concentrations much lower than those of O_3. Nevertheless, its diurnal pattern of occurrence is governed by photochemical processes (Fig. 2.3). At appropriate concentrations, ambient PAN can be toxic to plants.

To summarize, the dependency of gas-phase photochemical air pollution on various factors may be written as

$$\text{Photochemical air pollution} = \int \frac{(\text{NO}_x\ \text{conc.})(\text{organic compound conc.})(\text{radiation})(\text{temperature})}{(\text{wind speed})(\text{inversion / subsidence height})}$$

Notes 2.3. Production of peroxyacetyl nitrate, PAN in the troposphere
$NO_2 + R\bullet \rightarrow CH_3C(=O)OONO_2$
$CH_3C(=O)OONO_2$ = PAN; NO_2 = nitrogen dioxide; R• = free radical, $CH_3C(=O)OO$

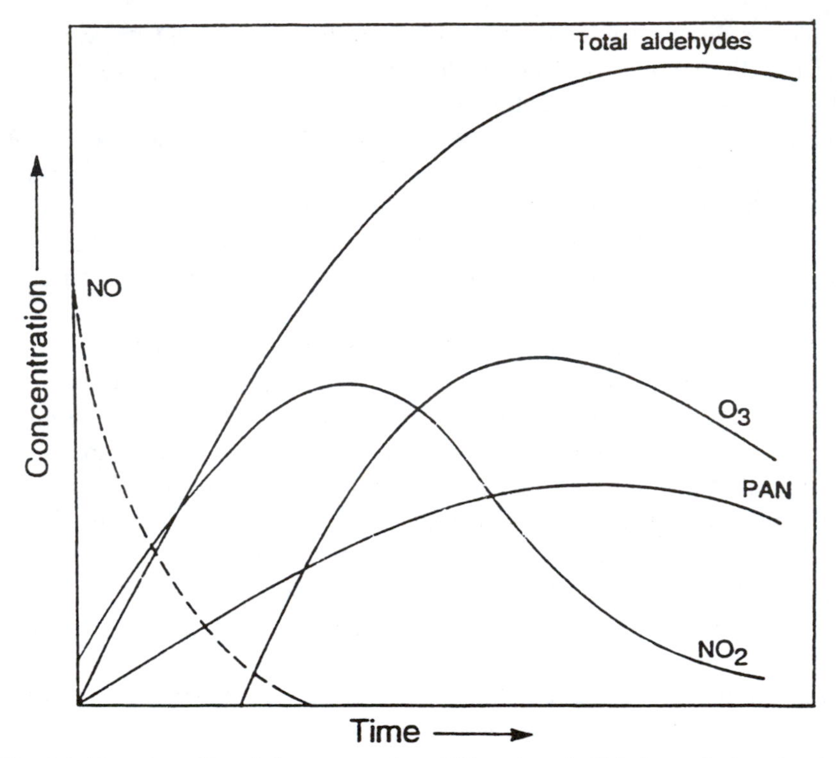

Fig. 2.3. Diurnal profiles of the reactants (e.g., NO_2 and total aldehydes or hydrocarbons) and products (e.g., O_3 and PAN) in photochemical air pollution.

Particulate Matter

As stated previously, particulate matter (aerosols) in the atmosphere can be either **primary** or **secondary** in origin and can be in either a dry or wet state. It is produced naturally (e.g., pollen, spores, salt spray, and soil erosion) and by human activity (e.g., soot, fly ash, and cement dust) (Table 2.7) and occurs in a broad range of sizes (Table 2.8). Nonetheless, on the basis of their atmospheric distribution, particles or aerosols are separated into three major categories: Aitken nuclei, fine particles, and coarse particles (Fig. 2.4).

Although a major fraction of the particulate matter in the atmosphere is composed of inorganic chemical constituents, organic pollutants can also be found in the particle phase (and in the gas and vapor phases). Among these are polyaromatic hydrocarbons (**PAH** or **POM**) (Fig. 2.5); the best known member of the group is benzo[*a*]pyrene (BaP). These compounds are derived from a six-carbon benzene ring, and several members of the group are carcinogens. This subject is discussed further in Chapter 6, Air Quality and Human Health. Table 2.9 provides a

Table 2.7. Source estimates of tropospheric particulate matter production[a]

Source	Weight (% of total)
Natural	
Primary	
Sea spray	28
Windblown dust	9.3
Forest fires	3.8
Volcanoes	0.09
Secondary	
Vegetation	28
Nitrogen cycle	14.8
Sulfur cycle	9.3
Volcanoes (gases)	0.009
Subtotal	94
Anthropogenic	
Primary	
Fossil fuel combustion and industrial dust	2.8
Agricultural cultivation	0.0009
Secondary	
Sulfates	2.8
Nitrates	0.56
Hydrocarbon vapors	0.065
Ammonia	0.028
Subtotal	6
Total	100

[a] Adapted from van de Vate and ten Brink, 1986.

Table 2.8. Typical size ranges for select types of particulate matter in the atmosphere

Particulate matter	Typical size range (μm)[a]
Fume	0.001–1.0
Mist	< 0.01–10.0
Carbon black	0.01–0.30
Tobacco smoke	0.01–1.0
Oil smoke	0.03–1.0
Ammonium sulfate	0.10–2.5
Sulfuric acid aerosol	0.10–2.5
Paint pigments	0.10–5.0
Insecticide dust	0.50–10.0
Dust	1.0–>300
Spray	10.0–>300
Spores	10.0–15.0
Pollen	10.0–100
Fine sand	12.0–200
Cement dust	15.0–100

[a] 1 μm = 10^{-6} meter.

summary of annual emission rates for some organic **(PAH** or **POM)** pollutants. At the United States and global scales, heating and power production, industrial processes, and coal-refuse burning are the three major sources of BaP. In comparison, at the global scale, forest and agricultural fires serve as the largest sources for total polycyclic organic matter (POM) (Table 2.9). For additional sources of organic air pollutants, refer to Table 2.2.

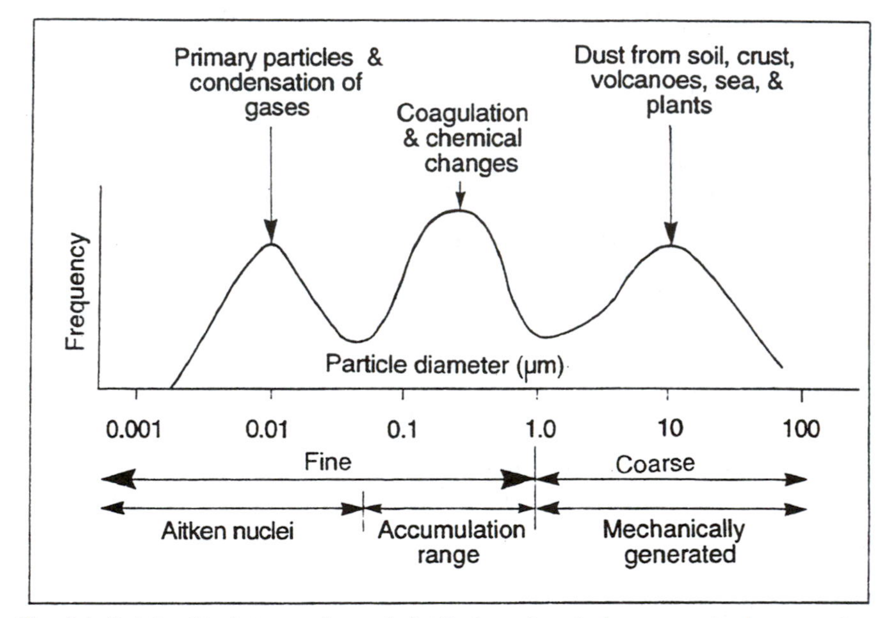

Fig. 2.4. Relationship between size and distribution of particulate matter in the atmosphere. (Adapted from Whitby, 1978)

Table 2.9. Estimated annual benzo[*a*]pyrene (BaP) and polycyclic organic matter (POM) emissions to the atmosphere[a,b]

Sources	BaP (kg × 10⁶) United States	BaP (kg × 10⁶) Global	POM global (kg × 10⁶)
Heating and power production	0.43 (36.8)[c]	2.36 (51.6)[c]	208.65 (4.8)
Industrial processes	0.18 (15.4)[d]	0.95 (20.8)[d]	...
Enclosed incineration	0.03 (2.6)	0.09 (2.0)	390.09 (9.0)
Coal-refuse burning	0.31 (26.5)	0.62 (13.6)	...
Forest and agricultural fires	0.13 (11.1)	0.38 (8.3)	3,256.78 (75.1)[e]
Other refuse burning	0.07 (6.0)	0.13 (2.8)	480.81 (11.1)
Trucks and buses	0.01 (0.8)	0.03 (0.7)	...
Automobiles	0.01 (0.8)	0.01 (0.2)	...
Total	1.17	4.57	4,336.33

[a] Adapted from Edwards, 1983.
[b] Values in parentheses are percentages of the totals.
[c] From coal combustion, 91%, and from wood combustion, 8%.
[d] From coke production, 99%.
[e] From agricultural burning, 60%, and from natural forest fires, 40%.

With the increasing interest in energy conservation, many home dwellers in severe cold-climate areas have been using wood-burning fireplaces as sources of heat during cold winter months. Table 2.10 provides a summary of the chemical composition (polyaromatic hydrocarbons) of particles from wood burning. Given the many thousands of home dwellers using wood as fuel, it serves as a significant source of organic pollutants in urban settings in the northern hemisphere.

While **coarse particles** (>2.5 μm in diameter) are generated mainly by mechanical processes and combustion (primary pollutants), **fine particles** (<2.5 μm in diameter) are produced in the atmosphere primarily through photochemical reactions (secondary pollutants) (Notes 2.4 and 2.5).

Sulfuric acid (H_2SO_4), initially formed as a vapor, condenses rapidly to form liquid droplets. In contrast, because of its higher saturation vapor

Notes 2.4. Formation of sulfuric and nitric acids in the troposphere

$$OH + SO_2 + M \rightarrow HSO_3 + M \quad (1)$$
$$HSO_3 + O_2 \rightarrow HO_2 + SO_3 \quad (2)$$
$$SO_3 + H_2O \rightarrow H_2SO_4 \quad (3)$$

Similarly, $OH + NO_2 + M \rightarrow HNO_3 + M$

H_2O = water; H_2SO_4 = sulfuric acid; HNO_3 = nitric acid; HO_2 = peroxy radical; HSO_3 = sulfurous radical; M = matter, a third body; NO_2 = nitrogen dioxide; O_2 = molecular oxygen; OH = hydroxyl radical; SO_2 = sulfur dioxide; SO_3 = sulfur trioxide.

Table 2.10. Some examples of PAH (polyaromatic hydrocarbon) levels in particles from wood burning[a,b]

PAH	Without catalytic combustor ($\mu g\ g^{-1}$)	With catalytic combustor ($\mu g\ g^{-1}$)	Remaining (%)
Anthracene, A	3,600	130	3.6
Benz[*a*]anthracene, BaA	740	55	7.4
Chrysene/triphenylene, Ch/Tr	770	67	8.7
Fluoranthene, F	2,300	190	8.2
Phenanthrene, Ph	7,500	480	6.4
Pyrene, P	2,100	160	7.6

[a] Adapted from Tan et al., 1992.

[b] Averaged results from duplicate analyses with a margin of error of ±30%. Many of these compounds can be carcinogenic if present at sufficient concentrations.

pressure, HNO_3 remains a vapor. Nevertheless, because of their extremely high reactivity, both H_2SO_4 and HNO_3 are neutralized by other chemical constituents in the atmosphere (e.g., ammonia, NH_3), leading to the production of fine particles with altered chemical composition. The extent of neutralization is dependent upon the residence time of the acid molecules, the atmospheric concentrations of the neutralizing molecules,

Notes 2.5. Neutralization of sulfuric and nitric acids in the troposphere
$H_2SO_4 + NH_3 \rightarrow \rightarrow \rightarrow (NH_4)_2SO_4$ $HNO_3 + NH_3 \rightleftarrows NH_4NO_3$ Equilibrium
H_2SO_4 = sulfuric acid; HNO_3 = nitric acid; NH_3 = ammonia; $(NH_4)_2SO_4$ = ammonium sulfate; NH_4NO_3 = ammonium nitrate.

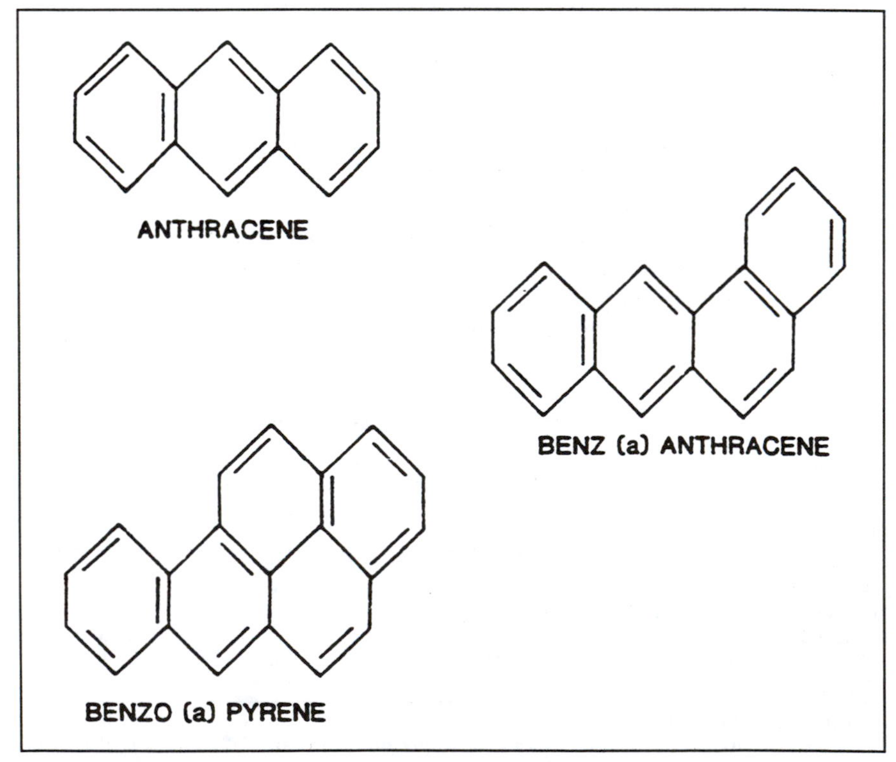

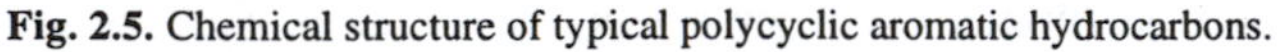
Fig. 2.5. Chemical structure of typical polycyclic aromatic hydrocarbons.

and the meteorological conditions. The implications of these characteristics and the features of deposition onto surfaces are discussed in Chapter 4, Transport of Air Pollution Across Regional, National, and International Boundaries.

Ambient Concentrations of Air Pollutants

In the United States, certain air pollutants are classified as **criteria pollutants**. This designation is based on the fact that these pollutants have documented effects on people, plants, or materials at concentrations, or approaching those, found in polluted air. In the United States, **National Ambient Air Quality Standards** (NAAQS) have been set for these pollutants "to protect public health and welfare" (Table 2.11). In Canada, there are national air-quality **objectives,** and in other countries,

Table 2.11. U.S. National Ambient Air Quality Standards (NAAQS) in effect in 1992

	Primary (health related)		Secondary (welfare related)	
Pollutant[a]	Type of average	Standard level concentration[b]	Type of average	Standard level concentration
CO	8-hr[c]	9 ppm (10 mg m^{-3})	No secondary standard	
	1-hr[c]	35 ppm (40 mg m^{-3})	No secondary standard	
NO_2	Annual arithmetic mean	0.053 ppm (100 µg m^{-3})	Same as primary standard	
O_3[f]	Maximum daily 1-hr average[d]	0.12 ppm (235 µg m^{-3})	Same as primary standard	
Pb	Maximum quarterly average	1.5 µg m^{-3}	Same as primary standard	
PM-10[f]	Annual arithmetic mean[e]	50 µg m^{-3}	Same as primary standard	
	24-hr[e]	150 µg m^{-3}	Same as primary standard	
SO_2	Annual arithmetic mean	80 µg m^{-3} (0.03 ppm)		
	24-hr[c]	365 µg m^{-3} (0.14 ppm)	3-hr[c]	1,300 µg m^{-3} (0.50 ppm)

[a] CO = carbon monoxide; NO_2 = nitrogen dioxide; O_3 = ozone; Pb = lead; PM-10 = particles <10 µm in diameter; and SO_2 = sulfur dioxide.

[b] Values in parentheses are approximate equivalent concentrations.

[c] Not to be exceeded more than once per year.

[d] The standard is attained when the expected number of days per calendar year with maximum hourly average concentrations above 0.12 ppm is ≤ 1, determined according to Appendix H of the Ozone NAAQS document.

[e] Particulate standards use PM-10 as the indicator pollutant. The annual standard is attained when the expected annual arithmetic mean concentration is ≤ 50 µg m^{-3}, and the 24-hr standard is attained when the expected number of days per calendar year above 150 µg m^{-3} is ≤ 1, determined according to Appendix K of the PM NAAQS document.

[f] In December 1996, the U.S. Environmental Protection Agency proposed revisions of the standards presented in this table. However, such revisions must be approved by the U.S. Congress before they can become law.

such as Germany, there are air-quality **guidelines**. In addition, the World Health Organization (WHO) has suggested desirable air-quality levels. The information on U.S. NAAQS is provided here to assist in understanding the data provided in Table 2.12 regarding present ambient concentrations of the criteria and other key air pollutants. A more detailed discussion of air-quality regulations is provided in Chapter 9, Control Strategies for Air Pollution.

Table 2.12 provides a comparison of ambient concentrations of several air pollutants in remote, rural, and moderately and heavily polluted geographic areas. As stated in Chapter 1, the designation of a geographic location as "remote" is highly questionable, since it is doubtful whether there are any areas on earth at the present time that have not been subjected to human influence. Nevertheless, "remote" areas are those locations far removed from human settlements. In examining and comparing the data between various types of locations presented in Table 2.12, the reader should keep in mind the broad geographic classification of air pollutants presented in Table 2.1.

Table 2.12. Typical peak concentrations of select pollutants observed in the troposphere over the continents[a]

	Type of atmosphere[c]			
Pollutant[b]	**Remote**	**Rural**	**Moderately polluted**	**Heavily polluted**
CO[d]	≤0.2 ppm	0.2–1 ppm	~1–10 ppm	10–50 ppm
HCHO	≤0.5–2 ppb	2–10 ppb	10–20 ppb	20–75 ppb
HNO_2	<30 ppt	0.03–0.8 ppb	0.8–2 ppb	2–8 ppb
HNO_3	≤0.03–0.1 ppb	~0.1–4 ppb	1–10 ppb	10–50 ppb
NH_3	15 ppt	1–10 ppb	1–10 ppb	10–100 ppb
NMHC[d]	≤65 ppbC	100–500 ppbC	300–1,500 ppbC	≥1.5 ppmC
NO	≤50 ppt	~0.05–20 ppb	0.02–1 ppm	~1–2 ppm
NO_2[d]	≤1 ppb	1–20 ppb	0.02–0.2 ppm	0.2–0.5 ppm
NO_3	≤5 ppt	5–10 ppt	10–100 ppt	100–430 ppt
O_3[d]	≤0.05 ppm	0.02–0.08 ppm	0.1–0.2 ppm	0.2–0.5 ppm
PAN	≤50 ppt	2 ppb	2–20 ppb	20–70 ppb
SO_2[d]	≤1 ppb	~1–30 ppb	0.03–0.2 ppm	0.2–2 ppm

[a] Adapted from Finlayson-Pitts and Pitts, 1986.

[b] CO = carbon monoxide; HCHO = formaldehyde; HNO_2 = nitrous acid; HNO_3 = nitric acid; NH_3 = ammonia; NMHC = non-methane hydrocarbon; NO = nitric oxide; NO_2 = nitrogen dioxide; NO_3 = nitrate; O_3 = ozone; PAN = peroxyacetyl nitrate; and SO_2 = sulfur dioxide. At present, the carbon dioxide **(CO_2)** concentration at the surface appears to be relatively homogeneous at ~360 ± 10 ppm. However, it is rising at the rate of ~0.6% per year.

[c] ppm = parts per million; **ppmC** = parts per million carbon equivalents; ppb = parts per billion; **ppbC** = parts per billion carbon equivalents; and ppt = parts per trillion.

[d] **Criteria pollutant.** This pollutant has documented effects on people, plants, or materials at concentrations, or approaching those, found in polluted air. In the United States, National Ambient Air Quality Standards (NAAQS) have been set "to protect public health and welfare." In addition to the pollutants listed in this table, there are two other criteria pollutants: total suspended particulate matter and lead.

In contrast to the gaseous pollutant concentrations presented in Table 2.12 (with the exception of nitrate, NO_3, which occurs as a vapor or as a particle), data on the ambient concentrations of particulate matter (**PM-10**, particles <10 μm in diameter) are presented in Table 2.13. The present 24-hr standard for PM-10 in the United States is **150 μg m^{-3}**. A problem with the standards for particulate matter is separating the contribution of human-made from natural sources. In some geographic locations, such as the United States' Great Plains, the contribution of natural sources (e.g., soil) can be highly significant. Nevertheless, PM-10 standards are mainly based on human health considerations, and in this context, technology is available to apportion or separate the contributions of natural versus human-made sources. Such approaches are based on source fingerprinting, in which certain elements and their concentrations are used as fingerprints for specific types of sources (e.g., lead for combustion of leaded gasoline or potassium for wood burning).

Table 2.13. Particulate matter (PM) loadings (24-hr geometric mean, μg m^{-3}) in selected urban areas, Global Environmental Monitoring System, 1979–1980[a,b]

Country (city)	PM[c]	Country (city)	PM[c]
Pakistan (Lahore)	640.1	Egypt (Cairo)	81.1
Kuwait (Kuwait)	521.7	Canada (Montreal)	77.9
Iraq (Baghdad)	503.8	Chile (Santiago)	77.8
India (Delhi)	450.9	Canada (Vancouver)	71.9
India (Calcutta)	423.6	Brazil (São Paolo)	71.5
Iran (Teheran)	357.9	Australia (Melbourne)	68.8
Indonesia (Jakarta)	260.1	Cuba (Havana)	60.4
Thailand (Bangkok)	221.7	**United States (New York City)**	**59.5**
Greece (Athens)	212.0	Japan (Osaka)	51.3
India (Bombay)	188.5	Poland (Worclaw)	50.7
Malaysia (Kuala Lumpur)	167.1	France (Gourdon)	43.3
United States (Los Angeles)	**150.8**[d]	Denmark (Copenhagen)	37.2
Spain (Madrid)	140.4	United Kingdom (London)	35.9
Romania (Bucharest)	140.2	United Kingdom (Glasglow)	34.2
Japan (Tokyo)	132.6	Poland (Warsaw)	31.7
Yugoslavia (Zagreb)	123.6	Peru (Lima)	28.9
United States (Chicago)	**118.5**	Ireland (Dublin)	25.2
Australia (Sydney)	110.6	Federal Republic of Germany (Frankfurt)	22.7
Philippines (Manila)	109.2	France (Toulouse)	19.5
Colombia (Bogota)	105.8	Belgium (Brussels)	15.5
Finland (Helsinki)	101.5	Colombia (Cali)	12.3
Romania (Craiova)	91.6	New Zealand (Christchurch)	11.7
Canada (Toronto)	90.9	New Zealand (Auckland)	4.0
Hong Kong	85.6		

[a] Adapted from World Health Organization, 1980. Locations in the United States are indicated in bold lettering.
[b] The data do not reflect current values for the loadings of particulate matter. Many cities in developed nations have reduced anthropogenic loadings substantially, while these values have gone up in major urban centers in the developing countries since 1980.
[c] Annual averages of selected sampling sites; 1 μg = 10^{-6} g.
[d] 1977–1978 data.

Table 2.14. Concentration ranges (ng m^{-3}) of some inorganic elements associated with particulate matter in the atmosphere[a,b]

Location	As	Cd	Hg
Remote	0.007–1.9	0.003–1.1	0.005–1.3
Rural	1.0–28	0.4–1,000	0.05–160
Urban			
Canada	7.7–626	2–103	<5
United States	2–2,320	0.2–7,000	0.58–458
Europe	5–330	0.4–260	0.1–5
Other	20–85	0.6–177	1.2–1.8
	Ni	**Pb**	**Se**
Remote	0.01–60	0.007–64	0.0056–0.19
Rural	0.6–78	2–1,700	0.01–3.0
Urban			
Canada	4–371	353–3,416	NA[c]
United States	1–328	30–9,627	0.2–30
Europe	0.3–1,400	10–9,000	0.01–127
Other	2.3–158	1.3–11,020	NA

[a] Adapted from Schroeder et al., 1987.
[b] One nanogram (ng) = 10^{-9} g. As = arsenic; Cd = cadmium; Hg = mercury; Ni = nickel; Pb = lead; and Se = selenium.
[c] Not available.

Table 2.15. Concentrations (ng m^{-3}) of some chemical constituents in **F**, fine (<2.5 μm in diameter) and **C**, coarse (>2.5 μm in diameter) particles at a rural (Ely) and a semiurban (Wright) site in Minnesota[a,b]

Site	Particle size	*n*	Al	Ca	Fe	NH_4
Ely	F	23	16 ± 5 (106)	18 ± 11 (257)	31 ± 10 (159)	632 ± 289 (5,863)
	C	23	52 ± 13 (234)	98 ± 27 (424)	101 ± 29 (643)	88 ± 79 (1,747)
Wright	F	28	43 ± 8 (201)	70 ± 17 (369)	56 ± 8 (170)	822 ± 383 (9,801)
	C	28	389 ± 69 (1,277)	820 ± 174 (3,946)	444 ± 67 (1,391)	44 ± 20 (417)
			NO_3	**Pb**	**Si**	**SO_4**
Ely	F	23	213 ± 30 (636)	9 ± 3 (62)	38 ± 21 (482)	2,293 ± 699 (11,242)
	C	23	150 ± 23 (418)	2 ± 1 (14)	197 ± 50 (958)	514 ± 203 (4,247)
Wright	F	28	261 ± 69 (1,796)	34 ± 5 (104)	112 ± 22 (492)	3,265 ± 1,876 (26,293)
	C	28	468 ± 111 (2,840)	5 ± 1 (28)	1,581 ± 281 (5,400)	357 ± 88 (2,730)

[a] Adapted from Pratt and Krupa, 1985.
[b] Values in parentheses = maxima; 1 ng = 10^{-9} g. Al = aluminum; Ca = calcium; Fe = iron; NH_4 = ammonium; NO_3 = nitrate; Pb = lead; Si = silica; and SO_4 = sulfate.

Although in a direct empirical sense, particulate matter (dust) can cause nasal problems such as sneezing, depending on the range of particle size, it can also lead to bronchial and other problems (see Chapter 6, Air Quality and Human Health). However, what is more important is the chemical composition of the particulate matter in reference to its toxicity. Table 2.14 provides a summary of concentration ranges of some important metals in the particulate phase. However, as previously stated, certain toxic trace metals and organic compounds can exist in both the particulate and gas phases. The extent of partition between the two phases can be specific to a particular type of molecule. For example, incineration of municipal refuse that includes plastic items will result in the production of hydrogen chloride (HCl) gas, which has the ability to force particulate mercury (Hg) to the gas phase. This is a critical consideration in any pollutant-emission control program (see Chapter 9, Control Strategies for Air Pollution). Other major problems are the complexity of the measurement technology, its cost, and the difficulty of using it to quantify the concentrations of atmospheric constituents that

Table 2.16. Typical concentration ranges for benzo[*a*]pyrenes associated with particulate matter in the atmosphere[a]

Location	Range ($\mu g\ m^{-3}$)[b]
Maritime	0.59–1.42
Nonurban continental	0.09–0.48
Urban continental	1.18–7.49

[a] Source: U.S. government documents.
[b] One microgram (μg) = 10^{-6} g.

Table 2.17. Concentrations of select organic compounds in air samples collected in Belgium[a]

Compound	Concentration ($ng\ m^{-3}$)[b] Particle phase (*P*)	Concentration ($ng\ m^{-3}$)[b] Gas phase (*G*)	Distribution factor (*P/G*)
Benz[*a*]anthracene and chrysene	12.2	3.87	3.15
Benzo[*a*]pyrene, benzo[*e*]pyrene and perylene	20.1	2.69	7.47
Benzo[*k*]fluoranthene and benzo[*b*]fluoranthene	23.1	2.01	11.5
Benzofluorenes	2.33	1.87	1.246
Fluoranthene	2.22	8.52	0.261
Methylphenanthrene and methylanthracene	0.90	10.2	0.088
Methylpyrene	0.93	. . .	. . .
Phenanthrene and anthracene	1.21	44.7	0.027
Pyrene	3.17	3.36	0.488

[a] Adapted from Cautreels and Van Cauwenberghe, 1978.
[b] One nanogram (ng) = 10^{-9} g.

occur in the particle-gas transition phase. Most regulatory agencies are not equipped to cope with these problems during routine analyses. Thus, frequently values reported through such efforts will likely be underestimates of the actual ambient levels. Nevertheless, such efforts, at least in a relative sense, can raise the flag for possible concern in a given situation.

Measurement of fine particles (<2.5 μm) can provide a reliable index of human contributions to PM-10 levels, since fine particles largely represent secondary pollutants formed in the atmosphere from precursor pollutants primarily generated by human activity (Fig. 2.4). Therefore, fine particles contain higher concentrations of many anthropogenic air pollutants such as ammonium (NH_4), sulfate (SO_4), and lead (Pb) (Table 2.15). In comparison, mechanically derived elements, for example, silica (Si), aluminum (Al), and calcium (Ca) from the soil, are found in higher concentrations in coarse particles. Normally, NO_3 (nitrate) is also found in higher concentrations in fine versus coarse particles (Table 2.15, Ely). However, NO_3 values for Wright (Table 2.15) appear to be exceptions, since the site is located in a sandy, agricultural area where farming practices appear to generate a significant fraction of nitrogen in the local air mass in the form of coarse particles.

As with Hg, many carcinogenic organic pollutants can occur in the particle and particle-gas transition phases (Tables 2.16 and 2.17). Similar to the pollutants described in Tables 2.12 and 2.14, ambient concentrations of organic pollutants such as BaPs vary from urban to nonurban areas (Table 2.16). Here, the proportion of particle to gas-phase concentrations is highly dependent on the structure of the chemical constituent, its molecular weight, and atmospheric conditions (Table 2.17). Because of many of these considerations and 1) the difficulties associated with measurements of their ambient concentrations, 2) present limitations in the knowledge of their geographic distributions, and 3) complexities in establishing exposure dose–human response relationships, with the exception of gross PM-10 and lead, **many particulate pollutants are not presently considered to be criteria pollutants** (i.e., there are currently no ambient air quality standards in the United States).

References

Cautreels, W., and Van Cauwenberghe, K. 1978. Experiments on the distribution of organic pollutants between airborne particulate matter and the corresponding gas phase. Atmos. Environ. 12:1133-1142.

Edwards, N. T. 1983. Polycyclic aromatic hydrocarbons (PAHs) in the terrestrial environment: A review. J. Environ. Qual. 12:427-441.

Finlayson-Pitts, B. J., and Pitts, J. N., Jr. 1986. Atmospheric Chemistry: Fundamentals and Experimental Techniques. John Wiley & Sons, New York.

Houghton, J. T., Jenkins, G. J., and Ephraums, J. J., eds. 1990. Climate Change: The IPCC [Inter-Governmental Panel on Climate Change] Scientific Assessment. Cambridge University Press, Cambridge.

Informatics, Inc. 1979. Air Quality Criteria for Carbon Monoxide. U.S. Environ. Prot. Agency Publ. EPA-600/8-79-022.

Lodge, J. P., Jr. 1969. Selections of The Smoake of London, Two Prophecies. Maxwell Reprint Co., Elmsford, NY.

Logan, J. A. 1983. Nitrogen oxides in the troposphere: Global and regional budgets. J. Geophys. Res. 88:785-807.

Logan, J. A. 1985. Tropospheric ozone: Seasonal behavior, trends, and anthropogenic influence. J. Geophys. Res. 90D6:10463-10482.

National Research Council. 1991. Rethinking the Ozone Problem in Urban and Regional Air Pollution. National Academy Press, Washington, DC.

Pratt, G. C., and Krupa, S. V. 1985. Aerosol chemistry in Minnesota and Wisconsin and its relation to rainfall chemistry. Atmos. Environ. 19:961-971.

Schroeder, W. H., Dobson, M., Kane, D. M., and Johnson, N. D. 1987. Toxic trace elements associated with airborne particulate matter: A review. J. Air Pollut. Control Assoc. 37:1267-1285.

Skelly, J. M., Krupa, S. V., and Chevone, B. I. 1979. Field surveys. Pages 12-1–12-30 in: Methodology for the Assessment of Air Pollution Effects on Vegetation. Proceedings of a Specialty Conference. W. W. Heck, S. V. Krupa, and S. N. Linzon, eds. Air Pollution Control Association, Pittsburgh.

Tan, Y. L., Quanci, J. F., Borys, R. D., and Quanci, M. J. 1992. Polycyclic aromatic hydrocarbons in smoke particles from wood and duff burning. Atmos. Environ. 26A:1177-1781.

van de Vate, J. F., and ten Brink, H. M. 1986. Sources of aerosols and acidic aerosols in the Netherlands: Not a special case. Pages 3-17 in: Aerosols: Research, Risk Assessment and Control Strategies. Proc. U.S.-Dutch Int. Symp., 2nd. S. D. Lee, T. Schneider, L. D. Grant, and P. J. Verkerk, eds. Lewis Publishers, Chelsea, MI.

Whitby, K. T. 1978. The physical characteristics of sulfur aerosols. Atmos. Environ. 12:135-159.

World Health Organization. 1980. Air Quality in Selected Urban Areas: 1977-1978. WHO Publ. 57.

Further Reading

Colbeck, I., and MacKenzie, A. R. 1994. Air Pollution by Photochemical Oxidants. Elsevier Scientific, Amsterdam.

Meszaros, E. 1981. Atmospheric Chemistry: Fundamental Aspects. Elsevier Scientific, Amsterdam.

Seinfeld, J. H. 1986. Atmospheric Chemistry and Physics of Air Pollution. John Wiley & Sons, New York.

CHAPTER 3

An Historical Perspective of Air Quality

> As soon as I had gotten out of the heavy air of Rome and from the stink of the smoky chimneys thereof, which, being stirred, poured forth whatever pestilential vapors and soot they had enclosed in them, I felt an alteration of my disposition.
>
> —Seneca; quoted by A. C. Stern et al.

Introduction

History can be described as representing a chronological record of significant events, usually including an explanation of their causes (*Webster's New Collegiate Dictionary*, 1993). Stern et al. (1973) in their book, *Fundamentals of Air Pollution,* provide an excellent summary of the historical aspects of air pollution. The following narrative is based on their treatment, and more information has been added. This should be a very enjoyable and educational analysis, since it provides an opportunity to examine the past up to the present. A view of the future is examined in Chapter 10, Education, Research, and Technology Transfer: An International Perspective.

Before the Industrial Revolution

One of the reasons the tribes of early history were nomadic was to periodically move away from the stench of the animal, vegetable, and human wastes they generated. When the tribesmen first learned to use fire, their living quarters were filled with the products of incomplete combustion because of the absence of proper ventilation. Examples of this can still be seen today in some of the more primitive parts of the world. After its invention, the chimney removed combustion products and cooking smells from the living quarters, but for centuries the open fire in the fireplace caused its emission to produce smoke.

Air pollution, associated with burning wood in Tutbury Castle in Nottingham, was considered "unendurable" by Eleanor of Aquitaine, the wife of King Henry II of England, and caused her to move in 1157. One hundred sixteen years later, coal burning was prohibited in London; and in 1306, Edward I issued a royal proclamation enjoining the use of "sea-coal" (mineral coal) in furnaces. Elizabeth I barred the burning of coal in London while Parliament was in session. The repeated necessity for such royal action would seem to indicate that coal continued to be burned despite these edicts. By 1661, the pollution in London had become bad enough to prompt John Evelyn to submit a brochure, "Fumifugium, or the Inconvenience of the Aer, and Smoake of London Dissipated (together with some remedies humbly proposed)," to King Charles II and Parliament. Recently, this brochure has been reprinted (Lodge, 1969), and it describes means of air pollution control that are still viable in the 20th century.

The principal industries associated with the production of air pollution during the centuries preceding the Industrial Revolution were metallurgy, ceramics, and the preservation of animal products. During the Bronze and Iron Ages, villages were often exposed to dusts and fumes from many sources. Native copper and gold were forged and clay was baked and glazed to form pottery and bricks before 4000 B.C. Iron was in common use and leather was tanned before 1000 B.C. Most of the methods of modern metallurgy were known before 1 A.D., although charcoal rather than coal or coke (the residue of coal or petroleum distillation) was used. Coal was mined and used for fuel before 1000 A.D.; however, it was not made into coke until about 1600. Coke was not used on a large scale in metallurgical practices until about 1700. These industries and their effluents as they existed before 1556 are best described in the book *De Re Metallica*, published by Georg Bauer, also known as Georgius Agricola. This book was translated into English in 1912 by Hoover and Hoover (see Agricola, 1556).

The Industrial Revolution

The Industrial Revolution was the consequence of harnessing steam to provide power to pump water and move machinery. The early years of the 18th century, during which Savery, Papin, and Newcomen designed their pumping engines, culminated in 1784 with Watt's development of a reciprocating engine, which reigned supreme until it was displaced by the steam turbine in the 20th century.

Steam engines and steam turbines require steam boilers, which until the advent of the nuclear reactor were fired by vegetable or fossil fuels. During most of the 19th century, coal was the principal fuel, although some oil was used for steam generation late in the century.

The predominant air pollution problem of the 19th century was smoke and ash generated from burning coal or oil in the boiler furnaces of stationary power plants, locomotives, and marine vessels and in fireplaces and furnaces used for home heating. Great Britain took the lead in addressing this problem.

Smoke and ash abatement in Great Britain was considered to be a health agency responsibility and was so confirmed by the first Public Health Act of 1848 and the later ones of 1866 and 1875. Air pollution from the emerging chemical industry was considered to be a separate matter and was made the responsibility of the Alkali Inspectorate created by the Alkali Act of 1863.

In the United States, smoke abatement (as air pollution control was then known) was considered a responsibility of municipalities. There were no federal or state smoke-abatement laws or regulations. The first municipal ordinances and regulations limiting the emission of black smoke and ash appeared during the 1880s and were directed toward industrial, locomotive, and marine sources rather than domestic ones. As the 19th century drew to a close, the air pollution in mill towns the world over had risen to a peak; damage to vegetation caused by the smelting of sulfide ores was recognized as a problem everywhere it was practiced.

The Twentieth Century

1900–1925

During the period from 1900 to 1925, there were great changes in both the technology leading to air pollution and its engineering-based control, but there were no significant changes in the legislation, regulations, understanding of the problem, or public attitudes toward the problem. As cities and factories grew, the pollution problem worsened.

One of the principal technological changes influencing air pollution was the replacement of the steam engine with the electric motor for operating machinery and pumping water. This change transferred smoke and ash emissions from the boiler house of the factory to the boiler house of the electricity-generating station. At the start of this period, coal was hand-fired in the boiler house. By the middle of the period, it was mechanically fired by stokers; by the end of the period, pulverized coal firing had begun to take over. Each form of firing added its own characteristic emissions to the atmosphere (Stern et al., 1973).

At the beginning of this period, steam locomotives came into the heart of the larger cities. By the end of the period, the urban terminals of many railroads had been electrified, thereby transferring much of the air pollution from the railroad right-of-way to the electricity-generating station. The replacement of coal by oil in many applications decreased ash emissions from those sources. There was rapid technological change in industry. However, the most significant change was the rapid increase in the number of automobiles from almost none at the turn of the century to millions by 1920 (Table 3.1).

The principal technological changes in the engineering control of air pollution were the perfection of the motor-driven fan, which allowed large-scale gas-treating systems to be built; the invention of the electrical precipitator, which made control of particulate matter in many processes feasible; and the development of a chemical engineering capability for the design of process equipment, which made control of vapor-phase effluents feasible.

1925–1950

It was during the period from 1925 to 1950 that present-day air pollution and solutions for its control emerged. Major air pollution episodes were recorded in Meuse Valley, Belgium (1930); Donora, Pennsylvania (1948); and Poza Rica, Mexico (1950) (Stern et al., 1973). Los Angeles smog first appeared during the 1940s. Major efforts to combat deleterious air pollution episodes were staged on many fronts. The Trail, British Columbia, smelter arbitration was completed in 1941 (Dean and Swain, 1944). The first National Air Pollution Symposium in the United States was held in Pasadena, California, in 1949 (Stern et al., 1973), and the first United States Technical Conference on Air Pollution was held in Washington, DC, in 1950 (McCabe, 1952). The first large-scale surveys

Table 3.1. Annual motor vehicle sales in the United States[a]

Year	Total	Year	Total
1900	4,192	1940	4,472,286
1910	187,000	1950	8,003,056
1920	2,227,347	1960	7,869,221
1930	3,362,820	1970	8,239,257

[a] Adapted from Stern et al., 1973. Data include factory sales for trucks, buses, and automobiles.

of air pollution were undertaken—Salt Lake City, Utah (1926); New York, New York (1937); and Leicester, England (1939); and the aforementioned episodes were intensively investigated (Stern et al., 1973). Air pollution research got a start in California. The technical foundation for air pollution meteorology was established in the search for means of disseminating, and protecting against, chemical, biological, and nuclear warfare agents. Toxicology came of age. All of these activities set the stage for the explosion of scientific research on air pollution that occurred during the second half of the 20th century.

A major technological change was the building of natural gas pipelines, which resulted in the rapid displacement of coal and oil as home heating fuel and consequently in a dramatic improvement in air quality. Today, one can clearly see the much-publicized decrease in the black smoke of Pittsburgh, Pennsylvania, and St. Louis, Missouri. The diesel locomotive began to displace the steam locomotive, thereby slowing the pace of railroad electrification. The internal combustion engine transport (bus) started its displacement of the electrified street car. The use of automobiles continued to proliferate (Table 3.1).

During this period, no significant national air pollution legislation or regulations were adopted anywhere in the world. However, the first state air pollution law in the United States was adopted by California in 1947.

1950–1970

In Great Britain, a major air pollution disaster hit London in 1952 (Ministry of Health, Mortality and Morbidity, 1954), resulting in the passage of the Clean Air Act in 1956 and an expansion of the authority of the Alkali Inspectorate.

The principal changes during this time were in the heating of homes. Previously, most heating was done by burning soft coal on grates in separate fireplaces in each room. A successful effort was made to substitute smokeless fuels for the soft coal used in this manner and central or electrical heating for fireplace heating. The outcome was a decrease in "smoke" concentrations, as measured by the blackness of paper filters. Particulate matter in the British air was transformed from 175 $\mu g\ m^{-3}$ in 1958 to 75 $\mu g\ m^{-3}$ in 1968 (Royal Commission on Environmental Pollution, 1971). In 1959, an international air pollution conference was held in London (Stern et al., 1973).

During these two decades, almost every country in Europe, as well as Japan, Australia, and New Zealand, experienced serious air pollution problems in the larger cities. As a result, during 1950–1970, several countries enacted their first national air pollution control legislation. By 1970, major national air pollution research centers had been set up in several countries (e.g., England, France, Japan, and the United States).

In the United States, the smog problem continued to worsen in Los Angeles and appeared in metropolitan areas all over the nation. In 1955, the first federal air pollution legislation was enacted, providing federal support for air pollution research, training, and technical assistance. Responsibility for the administration of the federal program was given to the Public Health Service (PHS) of the U.S. Department of Health, Education, and Welfare and remained there until 1970, when it was transferred to the newly created U.S. Environmental Protection Agency (EPA). The initial federal legislation was amended and extended several times between 1955 and 1970, greatly increasing federal authority, particularly in the area of control.

As in Europe, air pollution research activity expanded tremendously in the United Sates during these two decades. The headquarters of the federal research activity was at the Robert A. Taft Sanitary Engineering Center of the PHS in Cincinnati, Ohio, during the early years and was transferred to the research center of the EPA in Research Triangle Park, North Carolina, at the end of the period. An international air pollution congress was held in New York City in 1955 (Mallette, 1955), and the Second International Clean Air Congress was held in Washington, DC, in 1970 (Englund and Beery, 1971).

Technological interest during these 20 years focused on automotive air pollution and its control and on sulfur oxide pollution and its control by its removal from flue gases and fuel desulfurization. Air pollution meteorology came of age, and by 1970, mathematical models of atmospheric pollution were being energetically developed. A start was made in elucidating the photochemistry of air pollution. Air-quality monitoring systems became operational all over the world. A wide variety of measuring instruments became available.

The 1970s

The highlight of the 1970s was the emergence of the ecological or total environmental approach. Organizationally, this took the form of departments or ministries of the environment in governments at all levels throughout the world. In the United States, at the federal level, the EPA and in states such as Minnesota, counterpart organizations were charged with responsibility for air and water quality, noise abatement, etc. These efforts were paralleled in industry and in research and education.

The 1980s and Early 1990s

During these two decades, as accidents occurred at power plants at Three Mile Island in Pennsylvania in the United States (1979) and Chernobyl in the former U.S.S.R. (1986), public outcry forced Congress to stop the construction of new nuclear power plants in the United States. Meanwhile, environmentalists made the occurrence of acid rain and its environmental impacts a key national and international issue. The main cause of the acid rain phenomenon was identified as the emissions of sulfur and nitrogen oxides from coal-fired power plants. Canada took the United States to task for contributing to the acid rain in its eastern provinces; and similarly, Sweden and Norway blamed central Europe and Great Britain, respectively. In 1980, the U.S. Congress passed the Acid Precipitation Act and established the interagency U.S. National Acid Precipitation Assessment Program (NAPAP) to determine the impacts of acidic precipitation on the environment and develop strategies for its control. After millions of dollars in research support, NAPAP concluded that acidic precipitation was critical in adversely affecting only a small percentage of the lakes in the northeastern United States (U.S. NAPAP, 1990). Nevertheless, the Clean Air Act Amendment was passed by the U.S Congress in 1990 to regulate the emissions of sulfur and nitrogen oxides.

During the 1950s, a strange disease was observed on grapes in California and was identified as "stipple of grape." The cause was found to be nonbiological (i.e., not fungi, viruses, or bacteria). In the end, it was shown to be caused by photochemical smog. Here, there was an observed problem, the cause of which was later identified. In the case of acid rain, a cause or phenomenon was identified, and yet the extent of the problem in terms of significant geographic, ecological impact has not been clearly demonstrated (i.e., there are no large-scale crop or forest effects or deterioration of surface waters and lakes on a national or continental scale).

During the 1980s, as acid rain was becoming a national issue in the United States, widespread tree decline (Waldsterben) was causing a similar concern in the former West Germany and was being blamed on acid rain. Some hundreds of millions of dollars were spent on research in Germany alone. Other European countries and the United States followed this research trend. An international conference on "tree decline research" was held in Friedrichshafen, Germany, in 1989, and the German Minister for Science and Technology concluded that the German scientists did not know the definitive cause of the tree decline. Interestingly, while such massive research was in progress and since then, most trees in many geographic locations in Germany began to recover and exhibit normal morphology (see Chapter 8).

The 1980s witnessed a dramatic decline in the use of leaded gasoline in the United States and more recently in other countries such as Germany, with a corresponding increase in the use of cars equipped with catalytic converters (which use unleaded gasoline) and sales of smaller, more efficient cars. Sadly, however, this shift toward environmental concern did not transfer to urban centers such as Mexico City, Mexico. There, the increase in the use of automobiles and other traffic was uncontrolled as was the proliferation of photochemical smog. While smog levels in Los Angeles decreased because of air-quality regulations, levels continued to increase in urban centers such as Mexico City, Cairo, Egypt, and Bombay, India. While population growth, illiteracy, starvation, and economic considerations were critical issues in those countries, toxic metals and organic pollutants in the atmosphere became emerging issues in other countries such as the United States (see Chapter 10, Education, Research, and Technology Transfer: An International Perspective).

During the 1990s, "pollution prevention" has been the most advanced approach practiced by some major industries in the United States. This philosophy simply translates to "prevention" as better than "control." Such an approach involves major changes in process technology and the use of alternative and more efficient fuels for energy production. The concept of pollution prevention is perhaps the best representation of our civilization. **After all, cleanliness is what we desire, and it does not require an excuse** (e.g., acid rain and tree decline) **to practice**.

Although much of what occurred during the past relates to the direct effects of air quality on human health and welfare, the 1990s have become the decade of "global change." Global change includes a spectrum of issues: population growth, sociology, politics, and economics. It has also brought attention to the "greenhouse effect" and the potential for global warming. Air quality has been the focus of attention in these predicted climatic changes. Thus, the number of global climate change–environmental response models and their application has been growing at a rapid rate. Nevertheless, in viewing global climate change, one should keep in mind that Occam's razor, which lets many of us think that the simplest answer (e.g., the air temperature across the entire earth will simply increase) must be the right one, does not mean that nature must necessarily oblige in that manner. Integrative, multimedia analysis and control may provide the best promise for future environmental conservation, although such a philosophy will prove to be complex in its implementation.

The Fog of London and the Smog of Los Angeles and Mexico City: A Brief Analysis

London (Sulfurous) Fog

During the 13th century, coal began to replace wood for domestic heating and industrial uses in London; the impact of high-sulfur coal on air quality was dramatic. In his major 17th-century treatise, *Fumifugium* (see Lodge, 1969), John Evelyn wrote

> It is this horrid smoake which obscures our church and makes our palaces look old, which fouls our cloth and corrupts the waters, so as the very rain, and refreshing dews which fall in the several seasons, precipitate to impure vapour, which, with its black and tenacious quality, spots and contaminates whatever is exposed to it.

There was a renewed interest in air pollution and its health effects during the 20th century when a number of so-called killer fogs occurred. During these severe episodes, the numbers of deaths exceeded those expected for particular times of year and locations; during the most devastating incident, 4,000 fatalities were recorded in London in 1952. The conditions during these episodes tended to be characterized by heavy fogs and low-level inversions, which concentrated the pollutants in relatively small volumes. During the 1952 London incident, for example, Wilkins (1954), as cited by Finlayson-Pitts and Pitts (1986), described the inversion heights as being so low that occasionally the tops of the stacks of a power station at Battersea, which were 103 m high, could be seen above the fog/smog. In some locations, inversion heights as low as 46 m occurred occasionally, and visibility in the center of affected areas was less than 20 m.

The actual pollutants or combinations of pollutants responsible for the excessive numbers of deaths have not been identified, although in all cases there were increased levels of SO_2 and particulate matter in the presence of fog. During the 1952 London episode, concentrations of SO_2 as high as 1.3 ppm and total particles of 4.5 mg m^{-3} were recorded (Wilkins, 1954; cited by Finlayson-Pitts and Pitts, 1986).

Subsequent to the 1952 London episode, Britain passed a clean air act to reduce emissions. Although meteorological conditions similar to those of December 1952 occurred in 1956 and 1962, the number of excess deaths that occurred declined dramatically.

Because the 1952 episode was the most dramatic recorded to date in terms of health effects, this type of air pollution, characterized by high

SO_2 and particle concentrations in the presence of fog, has since been dubbed **London smog**. It is also frequently referred to as **sulfurous smog**. The term **smog**, in fact, is derived from a combination of the words **smoke** and **fog**.

Los Angeles (Photochemical) Smog

Injury to certain vegetable crops grown in Los Angeles County was first observed in 1944. From the nature of the injury and the relative susceptibilities of various plants, it was established by plant pathologists at the University of California, Riverside, that phytotoxic agents such as sulfur dioxide and fluorine compounds were not responsible (Middleton et al., 1950). Instead, they described a new "gas type injury" as well as another form of damage seen "only during periods when heavy air pollution is accompanied by fog particles which presumably contain pollutants of an as yet undetermined nature."

This key paper was followed by that of Haagen-Smit et al. (1952), who reproduced the plant-injury symptoms caused by ambient Los Angeles air pollution by exposing crops to synthetic atmospheres containing mixtures of ozone and olefins (straight-chain or branched hydrocarbons with one or more double bonds) or to illuminated mixtures of NO_2 and olefins.

Clearly, this was a new type of oxidizing air pollution, very different from that responsible for the previous severe air pollution episodes associated with London fog. Because it was first recognized in the Los Angeles area, it is frequently dubbed **Los Angeles smog**. However, during the summer, virtually all major cities today suffer from this phenomenon, including, for example, Mexico City and Tokyo. Furthermore, the term smog is misleading, because smoke and fog are not key components. The more appropriate term is **photochemical air pollution** (Finlayson-Pitts and Pitts, 1986), which is currently used except when historical connotations are relevant.

In their classic series of papers, Haagen-Smit et al. soon established that a major (but not the sole) pollutant was ozone formed during the reaction of hydrocarbons (HCs) and oxides of nitrogen (NO_x) in air in the presence of sunlight (see Finlayson-Pitts and Pitts, 1986).

Since hydrocarbons and NO_x are major constituents of the exhaust from motor vehicles and since Los Angeles has year-round intense sunlight and appropriate meteorological and geographical characteristics, in retrospect the reason that photochemical air pollution was first identified there is clear (Table 3.2).

Table 3.2. Comparison of the nature and deleterious properties of sulfurous pollutants and photochemical smog[a,b]

Characteristic	Sulfurous compounds	Photochemical smog
First recognized	Centuries ago	Mid-1940s
Place of origin of modern-day concern	London	Los Angeles
Primary pollutants	SO_2, sooty particles	Organic compounds, NO_x ($NO + NO_2$)
Secondary pollutants	H_2SO_4, aerosol sulfates, sulfonic acids, etc.	O_3, PAN, HNO_3, aldehydes, particulate nitrates, sulfates, etc.
Temperature	Low (≤35°F or 1.5°C)	High (≥75°F or 24°C)
Relative humidity	High, usually foggy	Low, usually hot and dry
Type of inversion	Radiation (ground)	Subsidence (overhead)
Time of air pollution peak concentrations	Early morning	Noon to evening
Major toxic pollutant	SO_2	O_3
Geographic magnitude of impact	SO_2: immediate vicinity of source regions	O_3: regional, interregional, and even continental
Nature of major environmental concern	SO_2: fine-particle SO_4, acidic precipitation	Global climate change
Background concentration	SO_2: ≤1 ppb	O_3: ≤40 ppb
Concentration in heavily polluted areas	SO_2: 0.2–2.0 ppm	O_3: 0.2–0.5 ppm
Nature of deposition	SO_2: dry, closer to sources; SO_4: wet, farther away from sources	O_3: dry only
Velocity of dry deposition	SO_2: 0.1–4.5 cm sec^{-1}	O_3: 0.1–2.1 cm sec^{-1}
Health effects	SO_2: With particulate matter, decrease in nasal mucous flow rate, increase in nasal flow resistance, chronic bronchitis, mortality/morbidity, etc.	Reversible eye irritation by smog (**not O_3**), chronic bronchitis, changes in lung function, cough, shortness of breath, headache, increase in susceptibility to disease, e.g., influenza
Sensitive plant species	SO_2: Alfalfa, barley, red clover, oats, aster, aspen, birch, jack pine	Alfalfa, barley, bean, red and white clover, grape, oats, potato, radish, soybean, tobacco, wheat, ash, aspen, black cherry, tulip poplar, white pine
Typical symptoms[c] on broad-leaved plants	SO_2: Interveinal chlorosis, necrosis	Upper surface chlorosis, bleaching, bronzing, flecking, stippling, unifacial and bifacial necrosis
Foliar accumulation of the pollutant	SO_2: yes (as sulfur)	O_3: no
Primary mechanism of foliar injury	SO_2: bisulfite (HSO_3^-) and sulfite (SO_3^-)	O_3: Free radicals (e.g., OH*)
Chronic effects on crops and forests	SO_2: known (e.g., in Europe)	O_3: known (e.g., in the United States)

[a] Adapted from Finlayson-Pitts and Pitts, 1986.

[b] H_2SO_4 = sulfuric acid; HNO_3 = nitric acid; NO = nitric oxide; NO_2 = nitrogen dioxide; NO_x = oxides of nitrogen; O_3 = ozone; PAN = peroxyacetyl nitrate; SO_2 = sulfur dioxide; and SO_4 = sulfate.

[c] Symptoms on conifers are similar in both cases: tip chlorosis or necrosis spreading downward. Mottling and banding can be observed with O_3. Chlorosis = yellowing; interveinal = between veins; necrosis = death of tissue; and stippling = groups of cells about the size of a pin or nail head accumulating brown to dark red pigmentation at several locations on the leaf surface.

Mexico City (Photochemical) Smog

Present-day Mexico City was established in 1325 A.D. by the Aztec Indians on the banks of the now dead Lake Texcoco. Over time, increasing population growth, urbanization, vehicular traffic (Table 3.3), and the number of diverse stationary sources of air pollution (Table 3.4) coupled with its topography and the resultant frequency of prevalent stagnant air masses have made Mexico City one of the most polluted areas in the world today (de Bauer and Krupa, 1990). Like many urban centers of the developing nations of the world, Mexico City suffers from an abundance of population at the poverty level, illiteracy, and varying degrees of starvation. Yet, it is interesting to note that in 1985, approximately one of every six citizens of Mexico had access to a motor vehicle. At the present time in the central part of Mexico City, the average speed of traffic is about 32 km per hour. According to Vizcaíno Murray (1975), during that year 60% of Mexico City's air pollution was contributed by mobile sources, and atmospheric inversions that did not break up until as late as 1400 hours occurred on 180 days. All these features and the location of Mexico City in a valley surrounded by mountains provide an ideal setting for the production of photochemical smog.

In 1937, visibility values (an indicator of air quality) for Mexico City were in the 15-km range; in 1966, they were most frequently in the 5-km

Table 3.3. Increase in population and the number of motor vehicles in Mexico City, Mexico, between 1940 and 1985[a]

Year	Population	Number of vehicles
1940	1,757,530	48,000
1960	4,870,876	248,000
1970	8,355,000	680,000
1980	15,000,000	2,500,000
1985 estimate	18,000,000	>3,000,000

[a] Adapted from de Bauer and Krupa, 1990.

Table 3.4. Brief summary of the increase in stationary sources of air pollution in Mexico City[a]

Year	Stationary sources
1877	Beginning of industrialization
1975	About 131,000 (36%) of the nation's estimated 350,000 stationary sources were located in the Valley of Mexico (approximately 0.5% of the total geographic area of the nation).
1984	About 171,000 (47%) of the nation's total stationary sources were located in the Valley of Mexico.

[a] Adapted from de Bauer and Krupa, 1990.

range; and at the present time, it is very common to record values of <2 km (de Bauer and Krupa, 1990).

de Bauer (1972) successfully used the production of injury symptoms on the leaves of sensitive higher plants as a mechanism for examining the relative air quality of Mexico City. She concluded that ozone (O_3) and peroxyacetyl nitrate (PAN) were the key phytotoxic components of Mexico City smog. More recently, Bravo et al. (1988) reported a significant increase in ambient surface O_3 concentrations over Mexico City during 1986 and 1987.

Paul Miller of the U.S. Forest Service and the University of California, Riverside, compared the characteristics of O_3 production in Mexico City with those in Los Angeles (Miller, 1993) (Table 3.5). The main difference between the two cities is that while high O_3 concentrations persist throughout the year in Mexico City, the winter season is a respite time for the O_3 levels in Los Angeles. In addressing the O_3 problem in the Valley of Mexico City, the national (Mexican) government is attempting to implement a control policy: e.g., vehicles with certain license plate numbers cannot be driven during certain days of the week. If strictly enforced, this approach can reduce emissions of NO_x and HC on a daily basis. In addition, this approach can promote the use of car pools and public transportation. This is definitely a start in reducing urban smog formation. However, in examining the aforementioned control strategy, it should be noted that wood and coal burning (also

Table 3.5. Comparison of surface-level O_3-related characteristics of Mexico City and the Los Angeles area[a]

Characteristic	Mexico City	Los Angeles
Latitude	19°25′N	34°00′N
Elevation	2,250 m MSL[b]	104 m MSL
Topography	Mountain valley	Ocean front, inland mountain range
Rainy period	Mainly summer	Mainly winter
Periods of high O_3 concentration	Summer, between rains, and even higher during winter	Summer
UV-A (320–400 nm)[c]	Relatively high	Lower than in Mexico City
Vegetation	No respite from O_3	Respite from O_3 during winter
Impact on natural ecosystems		
Area	Ajusco (>80 km downwind from Mexico City)	San Bernardino National Forest (>120 km downwind from Los Angeles)
Sensitive major tree species	*Pinus hartwegii*	*Pinus ponderosa*

[a] Adapted from Miller, 1993.
[b] Mean sea level.
[c] The peak wavelength for the photolysis of NO_2 is ~398 nm. $NO_2 + hv \rightarrow NO + O$; $O + O_2 + M \rightarrow O_3 + M$; hv = light energy; M = matter, a third body; NO = nitric oxide; NO_2 = nitrogen dioxide; O = singlet or atomic oxygen; O_2 = molecular oxygen; and O_3 = ozone.

sources of NO_x and HCs) have been on the increase with the increased migration of poor people from the rural areas into Mexico City looking for opportunities for survival. This population is characterized by illiteracy and substandard living conditions (e.g., huts) and food supply. This has a profound impact on socioeconomic segregation of the society of Mexico City and the environmental consequences. As noted in a recent local newspaper article from Mexico City, newly planted trees are stolen for fuel at a rate faster than they can be planted. This overall subject is discussed further in Chapter 9, Control Strategies for Air Pollution.

References

Agricola, G. 1556. De Re Metallica. Basel. Reprinted, 1950, Dover Publications, New York. (English translation and commentary by H. C. Hoover and L. H. Hoover, 1912, Mining Magazine, London.)

Bravo, H. A., Perrin, G. F., Sosa, E. R., and Torres, J. R. 1988. Incremento de la contaminación atmosferica por ozono en la zona metropolitana de la Ciudad de México. Rev. Ing. Ambiental. 1:8-14.

de Bauer, L. I. 1972. Uso de plantas indicadoras de aeropoloutos en la Ciudad de México. Agrociencia 9(D):139-141.

de Bauer, L. I., and Krupa, S. V. 1990. The Valley of Mexico: Summary of observational studies on its air quality and effects on vegetation. Environ. Pollut. 65:109-118.

Dean, R. S., and Swain, R. E. 1944. Report submitted to the Trail Smelter Arbitral Tribunal. U.S. Bur. Mines Bull. 453.

Englund, H., and Beery, W. T., eds. 1971. Proc. Int. Clean Air Congr., 2nd. Academic Press, New York.

Finlayson-Pitts, B. J., and Pitts, J. N., Jr. 1986. Atmospheric Chemistry: Fundamentals and Experimental Techniques. John Wiley & Sons, New York.

Haagen-Smit, A. J., Darley, E. F., Zaitlin, M., Hull, W., and Noble, W. 1952. Investigation on injury to plants from air pollution in the Los Angeles area. Plant Physiol. 27:18-34.

Lodge, J. P., Jr. 1969. Selections of the Smoake of London, Two Prophecies. Maxwell Reprint, Elmsford, NY.

Mallette, F. S., ed. 1955. Problems and Control of Air Pollution. Reinhold, New York.

McCabe, L. C., ed. 1952. Air Pollution: Proceedings of the U.S. Technical Conference on Air Pollution. McGraw-Hill, New York.

Middleton, J. T., Kendrick, J. B., Jr., and Schwalm, H. W. 1950. Injury to herbaceous plants by smog or air pollution. Plant Dis. Rep. 34:245-252.

Miller, P. R. 1993. Response of forests to ozone in a changing atmospheric environment. Angew. Bot. 67:42-46.

Ministry of Health, Mortality and Morbidity. 1954. Mortality during the London fog of December 1952. Rep. Public Health Related Subject 95. Her Majesty's Stationery Office, London.

National Acid Precipitation Assessment Program (NAPAP). 1990. Draft: Final Assessment Report, vols. 1-3. NAPAP, Washington, DC.

Royal Commission on Environmental Pollution. 1971. First Report (February). Command 4585. Her Majesty's Stationery Office, London.
Stern, A. C., Wohlers, H. C., Boubel, R. W., and Lowry, W. P. 1973. Fundamentals of Air Pollution. Academic Press, New York.
Vizcaíno Murray, F. 1975. La Contaminación en México. Fondo de Cultura Económica, México, D.F.

Further Reading

Brimblecombe, P. 1987. The Big Smoke: A History of Air Pollution in London Since Medieval Times. Methuen, London.
Singer, C., Holmyard, E. J., Hall, A. R., and Williams, T. I., eds. 1954-1958. A History of Technology, vols. 1-5. Oxford University Press, London.

CHAPTER 4

Transport of Air Pollution Across Regional, National, and International Boundaries

And now worse visions of a viler age
Loom through the darkness of the future's night
A sickening fog of smoke from British coal
Drops in a grimy pall upon the land
Befouls the vernal green and chokes to death
Each lovely shoot, drifts in low poisoned cloud.

—Henrik Ibsen, *Brand*

Introduction

As stated in previous chapters, once an air pollutant is emitted into the atmosphere, its lifetime is governed by **deposition** (both dry and wet), **transport,** and **chemical transformation** (primary to secondary pollutants). Dry deposition occurs whenever there is no precipitation and chemical constituents are transferred from the atmosphere onto surfaces (Table 4.1).

Gaseous pollutants such as sulfur dioxide (SO_2) and ozone (O_3) are deposited onto surfaces by **diffusion** (the gradual mixing of the molecules of two or more substances in the air as a result of random thermal motion). Diffusion is highly dependent on atmospheric conditions that govern air turbulence (e.g., air temperature) and by the surface (e.g., roughness). For stationary point sources, pollutant (e.g., SO_2) concentrations in the plume aloft are observed in high concentrations at the ground level only during surface-related inversion conditions (see Fig. 1.2). This is also true for secondary pollutants and subsidence (atmospheric inversion aloft) conditions. An important point here is that high concentrations of pollutants in the atmosphere do not necessarily mean high concentrations at the surface. This is because, as with SO_2, meteorological conditions leading to the occurrence of high concentrations of even secondary pollutants such as O_3 are not necessarily ideal for the optimum flux or exchange of O_3 from the atmosphere onto the surface. Therefore, calculations must be made between gas (both primary and secondary) concentrations in the air and gas concentrations at the surface to determine which exposures actually produced an effect. This is an area of complexity and controversy. Nevertheless, such calculations are available in the literature, although because of the aforementioned difficulty, they are not incorporated into most efforts to establish cause-and-effect relationships or develop air-quality regulations.

As opposed to gaseous pollutants, fine particles (<2.5 μm in diameter) are transported from the atmosphere onto surfaces through **Brownian motion** (the random motion of microscopic particles suspended in a liquid or gas caused by collision with molecules of the surrounding medium). As previously stated in Chapter 2, fine particles serve as the major source for most trace molecules and ammonium (NH_4^+), sulfate (SO_4^{2-}), and nitrate (NO_3^-) ions (see Table 2.15). Because of their physical characteristics, fine particles have a relatively low rate of transfer from the atmosphere to surfaces and therefore have a prolonged lifetime in the atmosphere. They can be of regional concern (e.g., reduced visibility), continental concern (e.g., sulfate or ammonium deposition through precipitation), or global concern (e.g., the 1993 summer weather modification, possibly caused by the eruption of Mount Piñatubo and the consequent fine-particle load in the atmosphere). Thus, secondary gaseous pollutants (e.g., O_3) and fine particles are of regional- or continental-scale geographic concern (see Table 2.1).

In contrast to fine particles, coarse particles (>2.5 μm in diameter) are deposited rapidly onto surfaces by **gravitational settling** and therefore are frequently of local concern. However, there are exceptions. Under

Table 4.1. Some chemical species relevant to dry deposition and their distribution among trace gas, fine particle, and coarse particle fractions[a,b]

Chemical	Trace gases	Fine particles	Coarse particles
Carbon species	VOCs	Organic acids	Graphitic compounds
Crustal materials	. . .	Some	Dominant
Nitrogen species	HNO_3	. . .	. . .
	NH_3	Most NH_4^+	Some NH_4^+
	NO_x	Most NO_3^-	Some NO_3^-
Oceanic materials	. . .	Some NaCl	Most NaCl
Oxidants	O_3	. . .	. . .
	H_2O_2	. . .	. . .
Sulfur species	SO_2	Most SO_4^{2-}	Some SO_4^{2-}
Trace metals	. . .	Dominant	Some

[a] Source: National Acid Precipitation Assessment Program, 1987.

[b] H_2O_2 = hydrogen peroxide; HNO_3 = nitric acid; NaCl = sodium chloride; NH_3 = ammonia; NH_4^+ = ammonium ion; NO_3^- = nitrate ion; NO_x = oxides of nitrogen; O_3 = ozone; SO_2 = sulfur dioxide; SO_4^{2-} = sulfate ion; and VOCs = volatile organic compounds.

appropriate conditions, regional-scale, coarse particles can be transported upward into the jet stream and across continents. For example, dust from the Sahara Desert is known to be transported and deposited in Florida and Georgia during summer rainfall. These examples represent massive meteorological phenomena. In most cases, however, local surface emissions of coarse particles result in their local deposition.

Figure 4.1 is a schematic diagram of the relationships between emissions, deposition, transport, and transformation of primary to secondary air pollutants, e.g., sulfur dioxide (SO_2) to sulfate (SO_4^{2-}) or nitrogen dioxide (NO_2) to ozone (O_3). The amount and importance of both dry and wet deposition vary in time and in geographic space. For example, closer to its sources, dry deposition of SO_2 is important, while much farther away, wet deposition of sulfate (SO_4^{2-}) is important.

Long-Range Transport of Ozone and Fine Particles

The persistence of secondary pollutants, such as fine-particle sulfate (SO_4^{2-}), in the atmosphere means they can be transported across different geographical boundaries: state (e.g., from Illinois into Wisconsin),

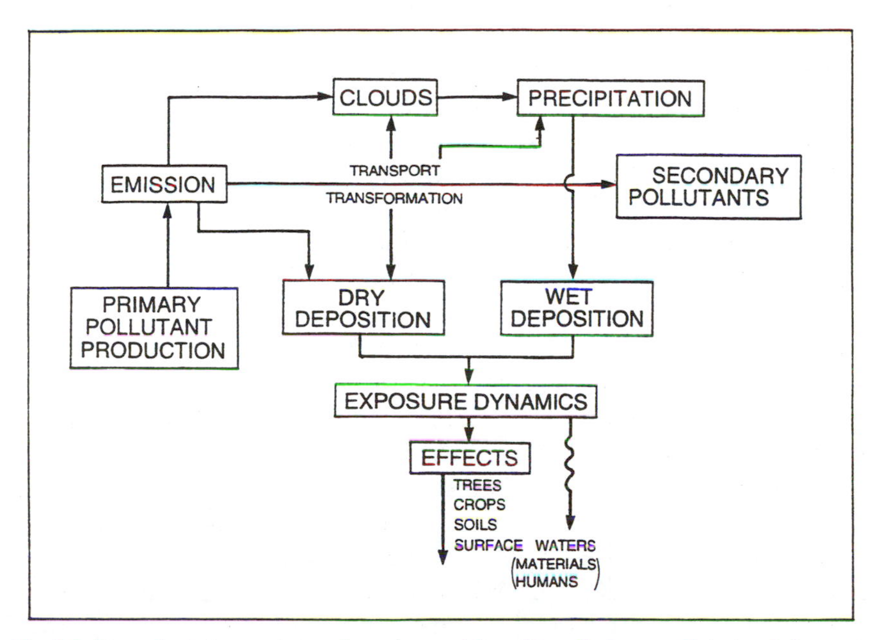

Fig. 4.1. Atmospheric transport, transformation, and deposition of primary pollutant emissions onto various environmental receptors resulting in effects. (Reprinted, by permission, from Legge and Krupa, 1990)

regional (e.g., Ohio River Valley into the northeastern United States), national (e.g., from the eastern United States into southeastern Canada), and international (e.g., across many nations within Europe). These phenomena are controlled by the density and strength of emissions of primary pollutants, prevailing meteorological conditions, transport and deposition processes, and the rate of transformation from primary to secondary pollutants. In this context, it is very important to mention that secondary pollutants such as ozone (O_3) do not represent the end of the line in their prolonged lifetime in the atmosphere and transport distance. O_3 is produced in the atmosphere and is highly reactive; therefore, its concentrations can be reduced even as it is being produced. If the production rate is greater than the consumption rate, then high concentrations occur. This is an important basis for the long-range transport of measurable O_3 concentrations.

In contrast to O_3, sulfuric acid (H_2SO_4) particles formed initially from sulfur dioxide (SO_2) are fine particles (<2.5 μm in diameter). Although H_2SO_4 is highly reactive in the atmosphere and can be converted into a salt, e.g., ammonium sulfate, $(NH_4)_2SO_4$, it remains a fine particle (<2.5 μm in diameter). As previously noted, these particles have the characteristic of Brownian motion, and thus their rate of dry deposition (when there is no precipitation) is relatively low compared with gases such as sulfur dioxide (SO_2) that are deposited by diffusion. Also, as described in Chapter 1, fine particles such as H_2SO_4 and $(NH_4)_2SO_4$ are very hygroscopic, **reduce visibility** on a regional scale, and serve as moisture condensation nuclei for cloud formation. These processes have implications in the **acid** or **acidic precipitation** phenomenon. This subject is discussed in detail in the following section.

Color Plate 1 provides an example of long-range transport of air pollutants and regional-scale processes. As previously stated, both fine-particle sulfate and O_3 are produced at the surface by reactions driven by solar radiation (photochemistry). Thus, high concentrations of these pollutants frequently occur together. Accumulation of fine-particle sulfate in the atmosphere leads to a reduction in visibility. Color Plate 1 also shows geographic contours of the number of miles (1 mile = 1.6 km) of visibility at various locations in the eastern and central United States at noon on June 29, 1975. In Minnesota, visibility was 4–8 miles (6.4–9.6 km), and in the clear area to the northwest, visibility was >15 miles (>18 km). For the same day, geographic contours of maximum hourly O_3 concentrations are provided. In the eastern half of Minnesota, maximum hourly O_3 concentrations were >70 ppb. During this episode, long-range air pollutant transport brought O_3 and fine-particle sulfate together into

Minnesota. Color Plate 1 also shows the path of the air parcels determined with weather-measurement data collected at 600 m above the ground at various geographic locations in the United States. The arrows indicate the location of the same air mass at 12-hr intervals, starting at 0 hours Greenwich mean time. The specific airflow patterns shown in Color Plate 1 were facilitated by the presence of a **high pressure system** in the northeastern Atlantic seaboard resulting in the **clockwise** airflow through the east central United States and Ohio River Valley into Minnesota. To place this overall phenomenon into perspective, it is interesting to note that **upwind** regions such as Ohio emit approximately 15 times more SO_2 (precursor for fine-particle sulfate) and Texas emits approximately seven times more NO_x (precursors for O_3) than Minnesota.

Acidic Precipitation (Rain)

The long-range transport of O_3 and fine-particle sulfate is very important on a large geographic scale, and so is the phenomenon of acidic precipitation (rain). The occurrence of acid rain caused much national and international concern from the mid-1970s through the 1980s because of its observed and/or predicted adverse effects on crops, forests, soils, surface waters (streams and lakes), and materials. This subject, as it relates to crops and forests, is discussed later in Chapters 7 and 8, respectively.

In 1971 at the World Clean Air Congress in Stockholm, Sweden, Bert Bolin and his Swedish co-workers presented a paper entitled "Transport of Air Pollution Across International Boundaries." Bolin et al. (1971) showed that the emissions of SO_2 from Germany and other central European nations were being transported across Europe and deposited in southern Sweden as acid rain. The term **acid rain** is used in the literature to describe **rainfall with a pH value less than 5.68** (the pH value of distilled water at 25°C in equilibrium with CO_2 at 1 atmospheric pressure or 760 mm of mercury). Similarly, Norwegian scientists brought attention to the transport and deposition of sulfur and nitrogen pollution from England into their own country. In the United States, Likens et al. (1972) described the occurrence of acid rain as a regional-scale problem in the northeast. In spite of these reports, it is important to note that acid rain is not a newly discovered phenomenon. Indeed, in 1692 in his book, *A General History of the Air*, Robert Boyle referred to "nitrous or salino-sulphureous spirits" in the air. Some 180 years later, a treatise on acid rain was published in England in 1872 by Robert Angus Smith; 20 years earlier he had analyzed rain near Manchester and noted three types of areas as one traveled from the city to the surrounding countryside:

> . . . that with carbonate of ammonia in the fields at a distance, that with sulphate of ammonia in the suburbs and that with sulphuric acid or acid sulphate, in the town.

These observations are very important and remain true in principle even today. The chemical composition of rainfall at a geographic location is dependent upon the path of transport of the rain cells arriving at that location from upwind source regions and the nature of the atmospheric composition of the regions in the path. For example, although SO_2 is initially converted into H_2SO_4, it can be neutralized if sufficient amounts of ammonia are present in the atmosphere (e.g., over agricultural areas, including animal feedlots). Nevertheless, there are inorganic ions that are important in precipitation chemistry (Table 4.2).

In understanding the amount of deposition of key ions and the chemical composition of rainfall at a given location, it is important to know the meteorological history of such rain events (Fig. 4.2). It should be noted that the two sections of Figure 4.2 present a generalized description of the processes governing winter (snow) and summer (rain) precipitation and there are always exceptions. Nevertheless, as these patterns relate to Minnesota, winter storms frequently originate in the north-northwestern United States (Fig. 4.2, top), areas with low emission levels of primary pollutants (SO_2 and NO_x) and relatively low urbanization, and travel across major agricultural areas (fallow during the

Table 4.2. Some inorganic ions important in precipitation chemistry[a,b]

Cations	Anions
Ca^{2+}	Cl^-
H^+	CO_3^{2-}
K^+	NO_3^-
Mg^{2+}	PO_4^{3-}
Na^+	SO_3^{2-}
NH_4^+	SO_4^{2-}

[a] Source: National Research Council, 1983.

[b] All ions are presented here in their completely dissociated states. The reader should note, however, that various states of partial dissociation are possible as well (e.g., HSO_3^- and HCO_3^-). As **ions**: Ca^{2+} = calcium; Cl^- = chloride; CO_3^{2-} = carbonate; H^+ = hydrogen; HCO_3^- = bicarbonate; HSO_3^- = bisulfite; K^+ = potassium; Mg^{2+} = magnesium; Na^+ = sodium; NH_4^+ = ammonium; NO_3^- = nitrate; PO_4^{3-} = phosphate; SO_3^{2-} = sulfite; and SO_4^{2-} = sulfate.

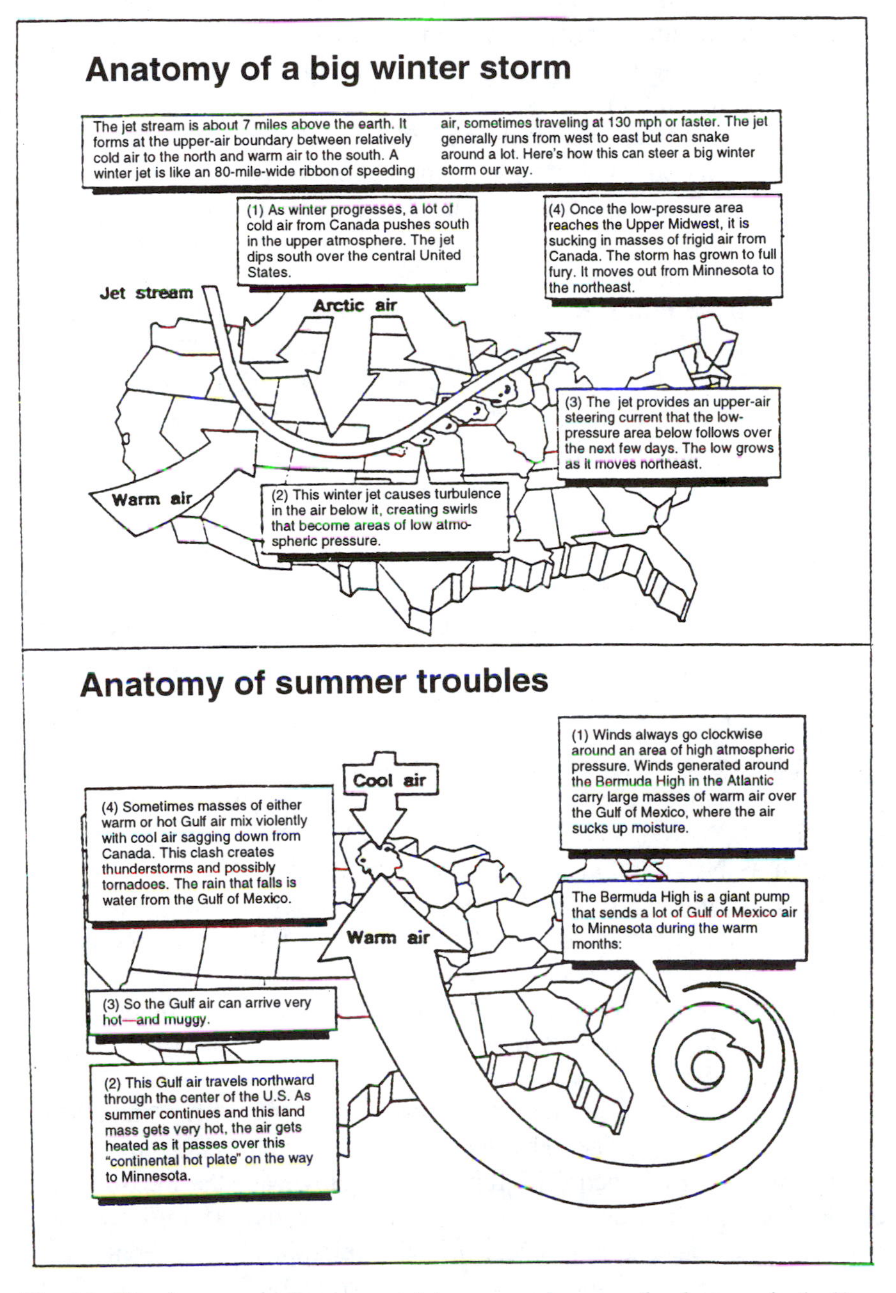

Fig. 4.2. Climatic processes that govern winter snow and summer thunderstorms in the Upper Midwest of the United States. (Source: National Weather Service)

winter). In contrast, summer rainfalls frequently originate from the Gulf of Mexico and travel across the central United States, a corridor of major metropolitan areas, industrial complexes (Ohio River Valley), and significant agricultural areas in the plains. The implications of these features in influencing the composition of precipitation are discussed in a later part of this section.

Figures 4.3 and 4.4 provide summaries of the annual median precipitation concentrations (mg/liter) of some major ions at various locations in the United States and the national ranking of each monitoring site. These data show a gradient starting in Minnesota, reaching the highest median SO_4^{2-} (sulfate) and NO_3^- (nitrate) concentrations in Ohio, Pennsylvania, and western New York, and decreasing into the northern New England states (Figs. 4.3C and 4.4C). The same pattern also holds true for H^+ (hydrogen ion) (Fig. 4.3A). In comparison, predominantly agricultural areas such as Minnesota and Nebraska rank at the top in the deposition of ammonium ion (Fig. 4.4B).

The basic concept governing the occurrence of acid rain is 1) SO_2 is converted to H_2SO_4, 2) NO_x is converted to HNO_3 (nitric acid), and these two acids reduce the pH level of rainfall to below 5.68. This overall concept has been the subject of some debate. Charlson and Rodhe (1982), Pratt et al. (1984), and Lefohn and Krupa (1988; based on the earlier observations of Krupa and Pratt, 1982) have questioned whether a pH value of 5.68 can be used as a natural background value for clean rainfall. While Charlson and Rodhe (1982) based their objection on the characteristics of natural bio-geochemical and hydrological cycles, Lefohn and Krupa (1988) based their argument upon the relationships between SO_4^{2-}, NO_3^-, and H^+ (Fig. 4.5). Sulfate, NO_3^-, and other major ions such as NH_4^+ and Ca^{2+} exhibit a curvilinear relationship with pH. Therefore, rainfall with the lowest concentrations of ions should be considered clean. The data presented in Figure 4.5 show that the lowest range of ion concentrations occur at pH values of about 4.8–5.3, perhaps a more accurate range for background pH values. The conclusion of Lefohn and Krupa is consistent with the independent results of Charlson and Rodhe (1982).

A second aspect of the debate concerns whether all SO_4^{2-} and NO_3^- in a rainfall are acids. The pH value of a solution represents the negative logarithm of its hydrogen (H^+) ion activity. However, the measurement of pH does not indicate what specific molecule the H^+ initially came from through dissociation. Therefore, the relationship between SO_4^{2-}, NO_3^-, and H^+ is established by statistical correlations (on a scale of 0 to +1, positive correlation, or −1, negative correlation). Such analyses show

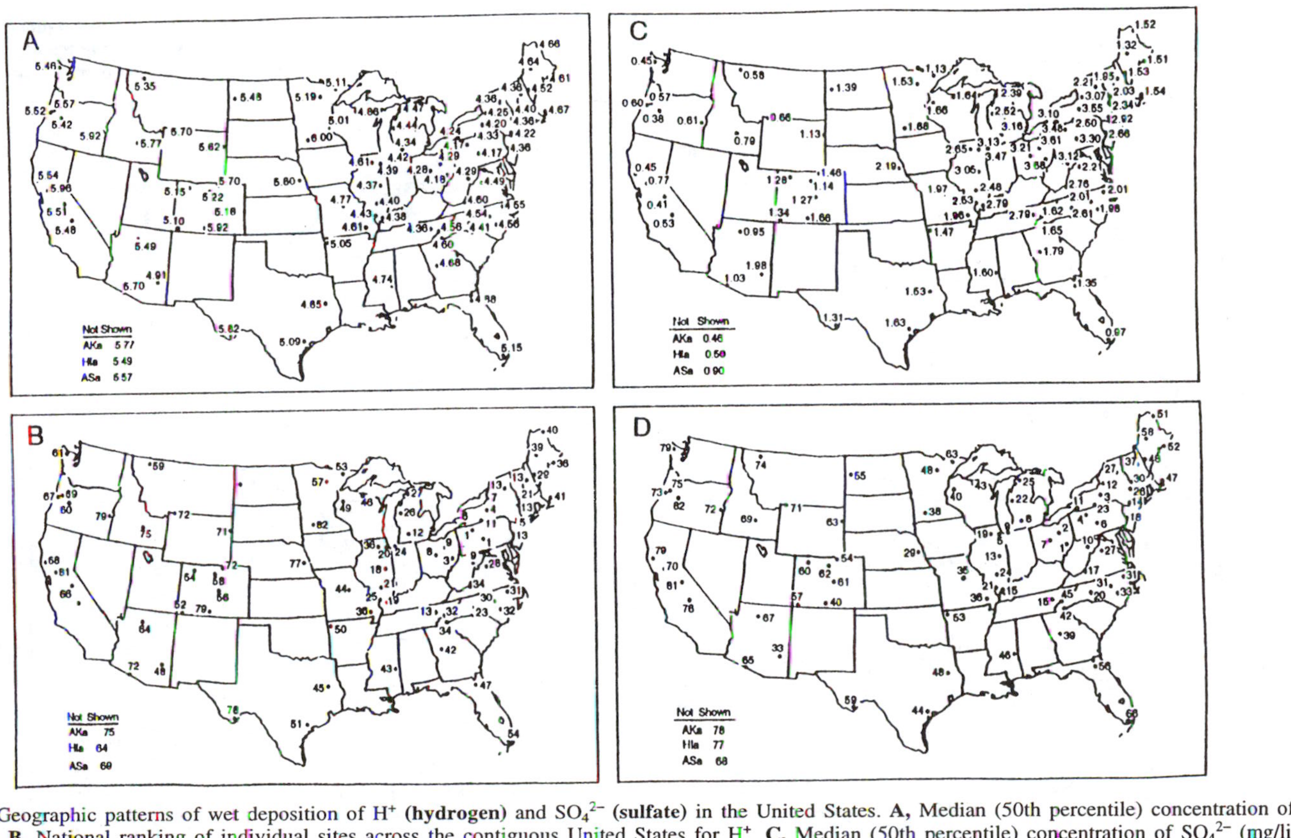

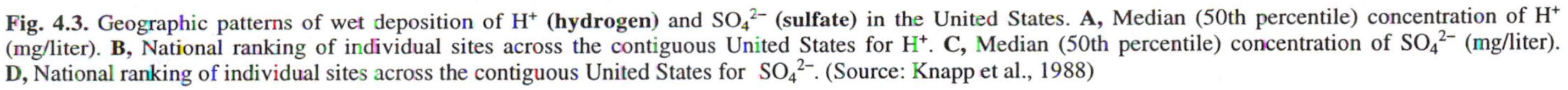

Fig. 4.3. Geographic patterns of wet deposition of H^+ **(hydrogen)** and SO_4^{2-} **(sulfate)** in the United States. **A,** Median (50th percentile) concentration of H^+ (mg/liter). **B,** National ranking of individual sites across the contiguous United States for H^+. **C,** Median (50th percentile) concentration of SO_4^{2-} (mg/liter). **D,** National ranking of individual sites across the contiguous United States for SO_4^{2-}. (Source: Knapp et al., 1988)

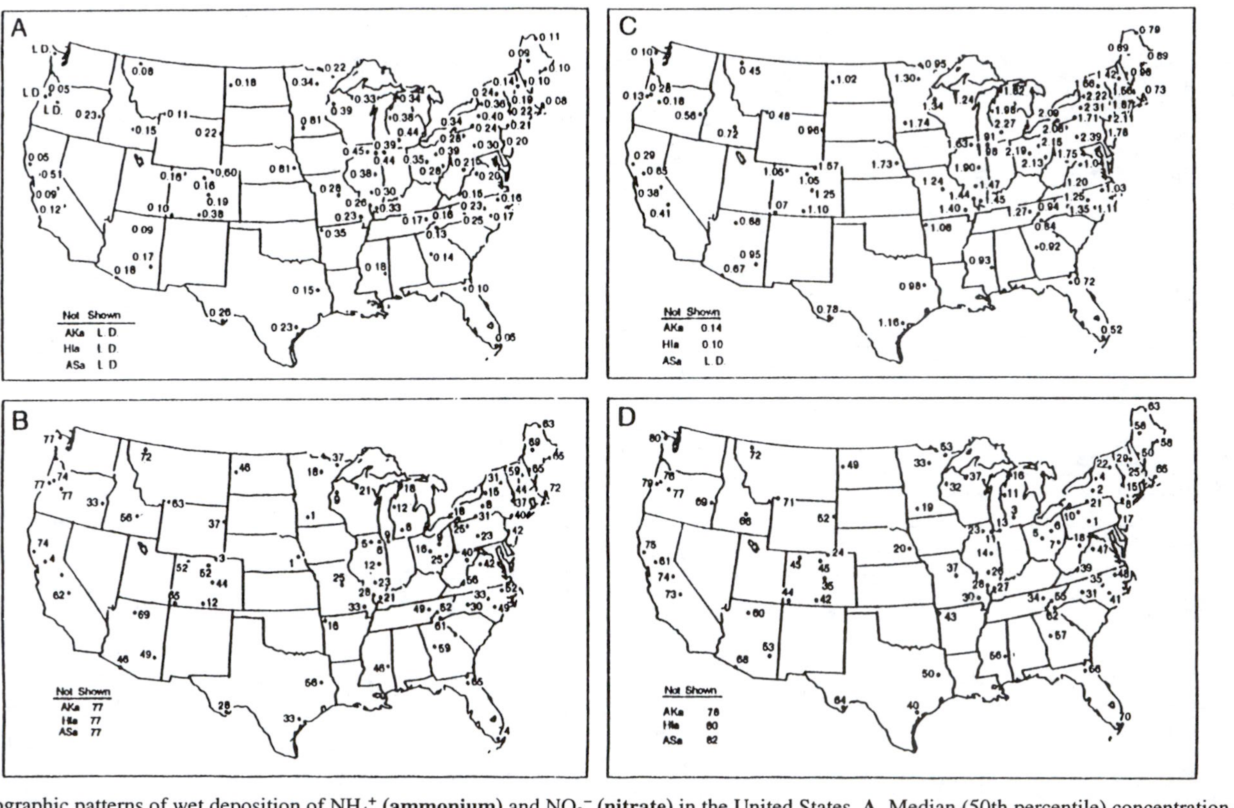

Fig. 4.4. Geographic patterns of wet deposition of NH_4^+ **(ammonium)** and NO_3^- **(nitrate)** in the United States. **A,** Median (50th percentile) concentration of NH_4^+ (mg/liter). **B,** National ranking of individual sites across the contiguous United States for NH_4^+. **C,** Median (50th percentile) concentration of NO_3^- (mg/liter). **D,** National ranking of individual sites across the contiguous United States for NO_3^-. (Source: Knapp et al., 1988)

that relationships between SO_4^{2-}, NO_3^-, and H^+ are highly variable in geographic space and time (e.g., rain event to event, month to month, or year to year). This is because, as previously stated, various fractions of the initially formed H_2SO_4 and HNO_3 may remain unchanged or may be neutralized by other chemical constituents in the atmosphere such as NH_3 (ammonia) or calcium (Ca^{2+}) before deposition. Therefore, the pH of rainfall is a product of the equilibrium between the acids and bases in solution.

Table 4.3 provides a comparison of the chemical composition of three different types of rainfall in Minnesota. **Type I** rainfalls contain relatively (a) high concentrations of H^+ (low pH); (b) high concentrations of NO_3^-; (c) high concentrations of SO_4^{2-}; and (d) low ratios of $NH_4^+:H^+$ (low degree of neutralization of acid, thus low pH). This type of rainfall is preceded by air trajectories coming through geographic areas (parts of the industrialized Ohio River Valley) with high SO_2 and NO_x emissions (Fig. 4.6A). **Type II** rainfalls are characterized by (a) intermediate H^+ concentrations and (b) low SO_4^{2-} and NO_3^- levels. These rainfalls are

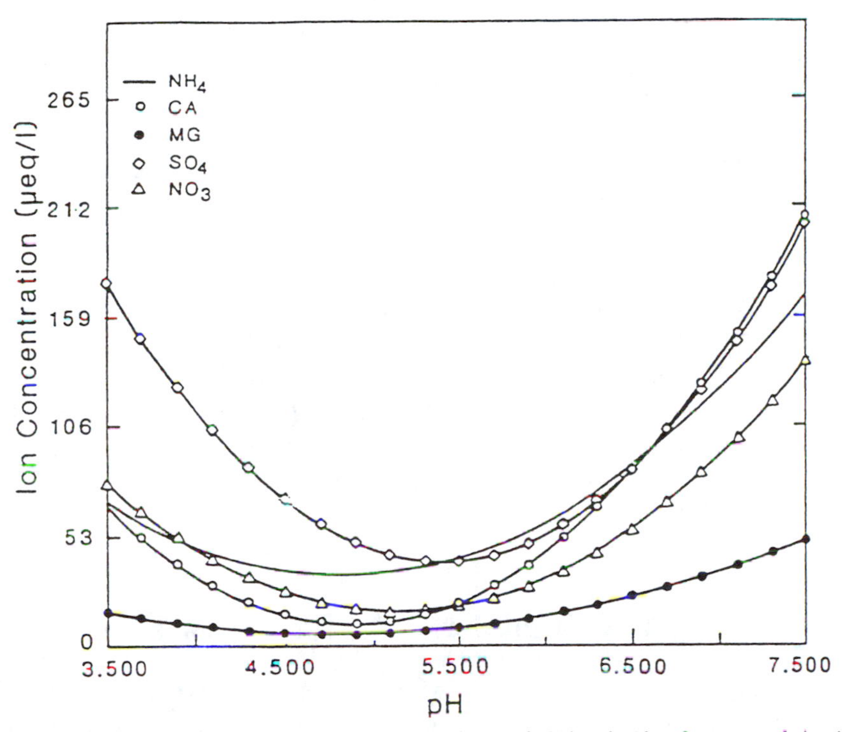

Fig. 4.5. Relationships between major ion concentrations and pH in rain (data from several sites in Minnesota and one site in Wisconsin were combined). (Source: Krupa and Pratt, 1982)

Table 4.3. Classification of daily rain samples in Minnesota, United States, by their relative chemistry[a]

Characteristic[b]	Type I	Type II	Type III
$\bar{x}$ H^+ (μeq/liter)[c]	63	36	2
$\bar{x}$ NH_4^+:H^+ (μeq/liter)[d]	1.44	0.76	168
$\bar{x}$ NO_3^- (μeq/liter)	54	21	64
$\bar{x}$ pH	4.20	4.44	5.80
$\bar{x}$ SO_4^{2-} (μeq/liter)	125	39	100

[a] Source: Krupa et al., 1987.
[b] As **ions**: H^+ = hydrogen; NH_4^+ = ammonium; NO_3^- = nitrate; and SO_4^{2-} = sulfate.
[c] eq = equivalence = molecular weight ÷ valence. This allows an ion balance of the amount of anions against cations and their mutual chemical reactivity.
[d] NH_4^+:H^+ is an indicator of the neutralization of H_2SO_4 (sulfuric acid) and HNO_3 (nitric acid). The higher the NH_4^+:H^+ value, the higher the neutralization of the strong acids, H_2SO_4 and HNO_3, by NH_3 (ammonia).

preceded by air trajectories coming from the north (Fig. 4.6B), areas of low emissions of SO_2 and NO_x, short travel distance, and low neutralization. In contrast to Types I and II, **Type III** rains contain (a) very low H^+ (high pH); (b) high concentrations of NO_3^- and SO_4^{2-}; and also (c) high neutralization (NH_4^+:H^+ ratio). These rainfalls are preceded by the typical air trajectories moving from the Gulf of Mexico and all the way through the central plains, regions of significant agriculture and a prevalence of feedlots (Fig. 4.6C).

This preceding discussion shows the complex relationships among geographic regions of primary pollutant emissions and their geographic path of transport; the varying degrees of atmospheric neutralization of the initially formed acids; and finally, the composition or nature of wet deposition at the sampling location.

The quality, relative concentrations of acids and bases, and frequency of occurrence of rain with varying chemical composition are all important considerations. For example, Type I rainfalls in Minnesota may have a biological significance different from that of Type III rainfalls. This is a subject that has not been well researched, particularly in the context of the occurrence of other air pollutants (e.g., O_3) and their joint effects.

A Brief Analysis of the Acidic Rain Phenomenon

Compared with sea water, rain is a highly dilute, poorly buffered mixture of chemical constituents. At most locations, rainfall is acidic in nature. The use of a pH value of 5.68 as the background is questionable,

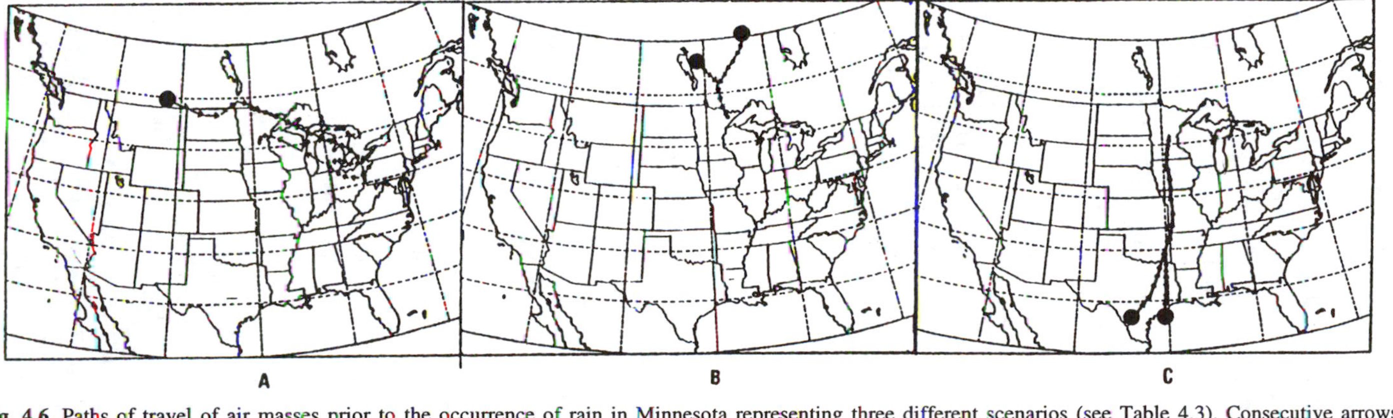

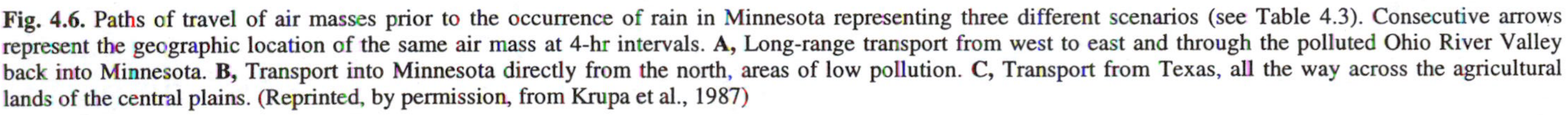

Fig. 4.6. Paths of travel of air masses prior to the occurrence of rain in Minnesota representing three different scenarios (see Table 4.3). Consecutive arrows represent the geographic location of the same air mass at 4-hr intervals. **A,** Long-range transport from west to east and through the polluted Ohio River Valley back into Minnesota. **B,** Transport into Minnesota directly from the north, areas of low pollution. **C,** Transport from Texas, all the way across the agricultural lands of the central plains. (Reprinted, by permission, from Krupa et al., 1987)

since the pH of rainfall with the lowest concentrations of many ions appears to be already in the range of 4.8 to 5.3. Nevertheless, SO_4^{2-} and NO_3^- contribute in various degrees to the acidity of rainfall, depending on the geographic location and time. In some locations such as the Upper Midwest, organic acids (e.g., formic acid, possibly formed through photochemical processes) can also contribute to the rainfall pH. The composition of snow, because it has only about one-tenth water content by its mass, has not been as thoroughly studied as that of rain. Presently, there are no instruments available that can collect snow satisfactorily and reliably as there are for rain. An additional consideration is the difference between the months when plants grow and the season when snow occurs. In surface water studies, however, the important role of snow, its spring thaw, and moisture runoff have been documented. The impacts of acidic precipitation on crops and forests constitute an area of debate (Chapters 7 and 8). Perhaps the adverse effects of acidic rain are more clearly documented for surface waters and materials (e.g., historical monuments and statues).

References

Bolin, B., Granat, L., Ingelstam, L., Johannesson, M., Mattsson, E., Odén, S., Rodhe, H., and Tamm, C. O. 1971. Air Pollution Across National Boundaries. The Impact on the Environment of Sulfur in Air and Precipitation. Sweden's Case Study for the United Nations Conference on the Human Environment. Royal Ministry for Foreign Affairs, Royal Ministry of Agriculture, Stockholm.

Boyle, R. 1692. A General History of the Air. Awnsham and John Churchill, London.

Charlson, R. J., and Rodhe, H. 1982. Factors controlling the acidity of natural rainwater. Nature 295:683-685.

Ibsen, H. 1978. Brand by Henrik Ibsen; A version for the English stage by Geoffrey Hill. Heinemann, London.

Knapp, W. W., Bowersox, V. C., Chevone, B. I., Krupa, S. V., Lynch, J. A., and McFee, W. W. 1988. Precipitation chemistry in the United States: Summary of ion concentration variability 1978–1984. Tech. Bull. Water Resources Research Institute, Cornell University, Ithaca, NY.

Krupa, S. V., and Pratt, G. C. 1982. Rainfall and Aerosol Chemistry in Minnesota and Wisconsin (1982). Minnesota/Wisconsin Power Suppliers Group. Northern States Power Company, Minneapolis. pp. 1-111.

Krupa, S. V., Pratt, G. C., and Teng, P. S. 1982. Air pollution: An important issue in plant health. Plant Dis. 66:429-434.

Krupa, S. V., Lodge, J. P., Jr., Nosal, M., and McVehil, G. E. 1987. Characteristics of aerosol and rain chemistry in north central U.S.A. Pages 121-128 in: Acidic Rain: Scientific and Technical Advances. R. Perry, R. M. Harrison, J. N. B. Bell, and J. N. Lester, eds. Selper, London.

Lefohn, A. S., and Krupa, S. V. 1988. The relationship between hydrogen and sulfate ions in precipitation—A numerical analysis of rain and snowfall chemistry. Environ. Pollut. 49:289-311.

Legge, A. H., and Krupa, S. V., eds. 1990. Acidic Deposition: Sulphur and Nitrogen Oxides. Lewis Publishers, Chelsea, MI.

Likens, G. E., Bornmann, F. H., and Johnson, N. M. 1972. Acid rain. Environment 14:33-40.

National Acid Precipitation Assessment Program (NAPAP). 1987. Interim Assessment Report, vols. 1-4. NAPAP, Washington, DC.

National Research Council. 1983. Acidic Deposition. Atmospheric Processes in Eastern North America. National Academy Press, Washington, DC.

Pratt, G. C., Coscio, M. R., and Krupa, S. V. 1984. Regional rainfall chemistry in Minnesota and west central Wisconsin. Atmos. Environ. 18:173-182.

Smith, R. A. 1872. Air and Rain. Longmans, Green, London.

Further Reading

Cowling, E. B. 1982. Acid precipitation in historical perspective. Environ. Sci. Technol. 16:110A-113A.

Hidy, G. M. 1984. Source-receptor relationships for acid deposition: Pure and simple? J. Air Pollut. Control Assoc. 34:518-531.

Schwartz, S. E. 1989. Acid deposition: Unraveling a regional phenomenon. Science 243:753-762.

CHAPTER 5

Air Pollution and Global Climate Change

> Little upon his or her little earth, a person contemplates the universe of which he or she is both judge and victim.
>
> —W. H. Auden, *Commentary*

Introduction

"Global climate change," the "**greenhouse effect**," and "global warming" are all terms that express the concerns of the public, the scientific community, and governments on the international scale. All of these terms describe the observed or predicted changes in the interactions between the chemical and physical climate of the earth.

The popular perception of this subject was stated by Robert A. Rankin in the *St. Paul Pioneer Press Dispatch*, December 4, 1988, as follows:

> Sunlight strikes the earth, heating the rock and water of the surface. The earth then radiates the heat as infrared rays. An equilibrium is thus established between the solar energy received and the heating of the earth and atmosphere. Carbon dioxide and other gases are released into the atmosphere from natural sources, such as plant and animal life, and artificial sources, such as factories and cars. The atmosphere is composed primarily of nitrogen, 78%, and oxygen, 21%, with other trace gases such as carbon dioxide, argon, hydrogen, and helium contributing minute amounts.
>
> Gases accumulate in the atmosphere and act like glass in a greenhouse, letting in the warming rays, but inhibiting the escape of infrared rays [Fig. 5.1].
>
> Scientists know a lot less about the greenhouse effect than the news media may have led you to believe during the long, hot summer.
>
> To be sure, there is no debate among atmospheric scientists that a greenhouse effect exists. It is a fact of nature, it is getting worse and it almost certainly will cause the earth's climate to warm-up.
>
> But warm-up how much? How fast? With what impact? On those critical questions, scientists disagree.

Traditionally, the effects of climate on plant growth and productivity have been viewed in the context of physical climatic parameters such as temperature and precipitation by agricultural and forest meteorologists. While plant physiologists have studied the assimilation of carbon dioxide (CO_2) by plants, others have examined the adverse effects of trace gas air pollutants such as ozone (O_3) and sulfur dioxide (SO_2) on plant growth and productivity. In the most recent years, however, the compartmentalized view of climate-plant interactions is changing as our understanding of the dynamic interactions between the chemical and physical climate is rapidly increasing.

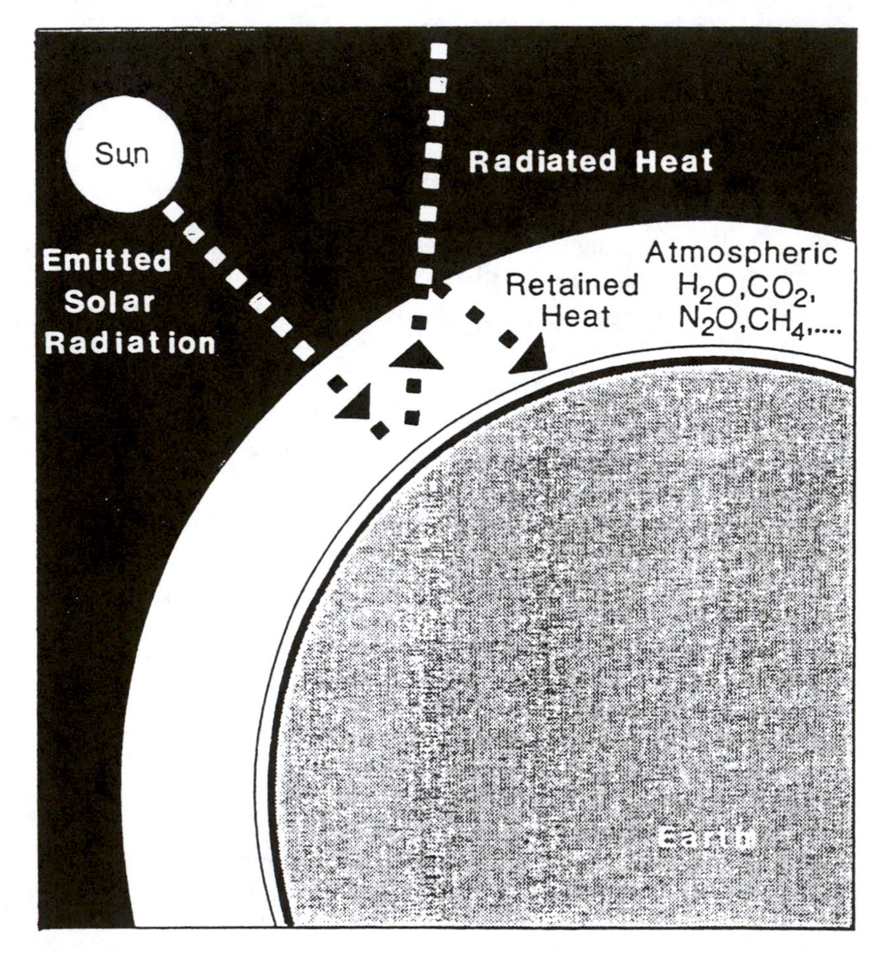

Fig. 5.1. The **natural greenhouse effect** in the earth's atmosphere. CH_4 = methane; CO_2 = carbon dioxide; H_2O = water; and N_2O = nitrous oxide.

Processes and Products in Global Climate Change

Any climate modification and global warming are governed by interactions between tropospheric and stratospheric processes (Wuebbles et al., 1989). The key atmospheric constituent in these interactions is O_3.

As stated in Chapter 1, atmospheric O_3 is produced by reactions that are natural and by those that are governed by human activity (Krupa and Manning, 1988). By volume, atmospheric O_3 content varies with altitude above the earth's surface (see Fig. 1.1) roughly as follows: 10% at 0–10 km, 80% at 10–35 km, and 10% above 35 km (Cicerone, 1987). Ozone concentrations at the surface also fluctuate with latitude, and this variation is strongly influenced by urbanization and industrialization.

As described in Chapter 1, a series of photochemical reactions involving O_3 and molecular oxygen, O_2, occurs in the stratosphere. Ozone strongly absorbs solar radiation in the region from $\approx$210 to 290 nm, whereas O_2 absorbs radiation at $\lesssim$200 nm. The absorption of light primarily by O_3 is a major factor in the increase in temperature with altitude in the stratosphere. Excited O_2 and O_3 photodissociate, initiating a series of reactions in which O_3 is both formed and destroyed, leading to a steady-state concentration of O_3 (see Notes 1.1 and Cicerone, 1987). This O_3 serves as a shield that keeps biologically harmful solar ultraviolet (UV) radiation from reaching the earth's surface. It also initiates key stratospheric chemical reactions and transforms solar radiation into heat, and the subsequent differences in the air pressure induce the mechanical energy of atmospheric winds. The flux of photochemically active UV-B (280–320 nm) photons into the troposphere is limited largely by the amount of stratospheric and tropospheric O_3 (see Table 1.2). In addition to this protective stratospheric O_3, clouds also reflect a large part of the incoming solar radiation, causing the **albedo** (the fraction of incident electromagnetic radiation reflected by a surface) of the entire earth to be about twice what it would be in the absence of clouds. Clouds cover about one-half the earth's surface, doubling the proportion of sunlight reflected back into space by up to 30% (Fig. 5.2).

Ever since the publications by Johnston (1971) and Molina and Rowland (1974), human activities have been seen as causing substantial depletion of the stratospheric O_3 layer by increasing the global concentrations of certain atmospheric chemicals such as methane (CH_4), nitrous oxide (N_2O), methyl chloride (CH_3Cl), synthetic chlorofluorocarbons (CFCs), chlorocarbons (CCs), and organo-bromines (OBs) (Table 5.1).

Reduced amounts of atmospheric O_3 will permit disproportionately large amounts of UV-B radiation to penetrate to the earth's surface. For

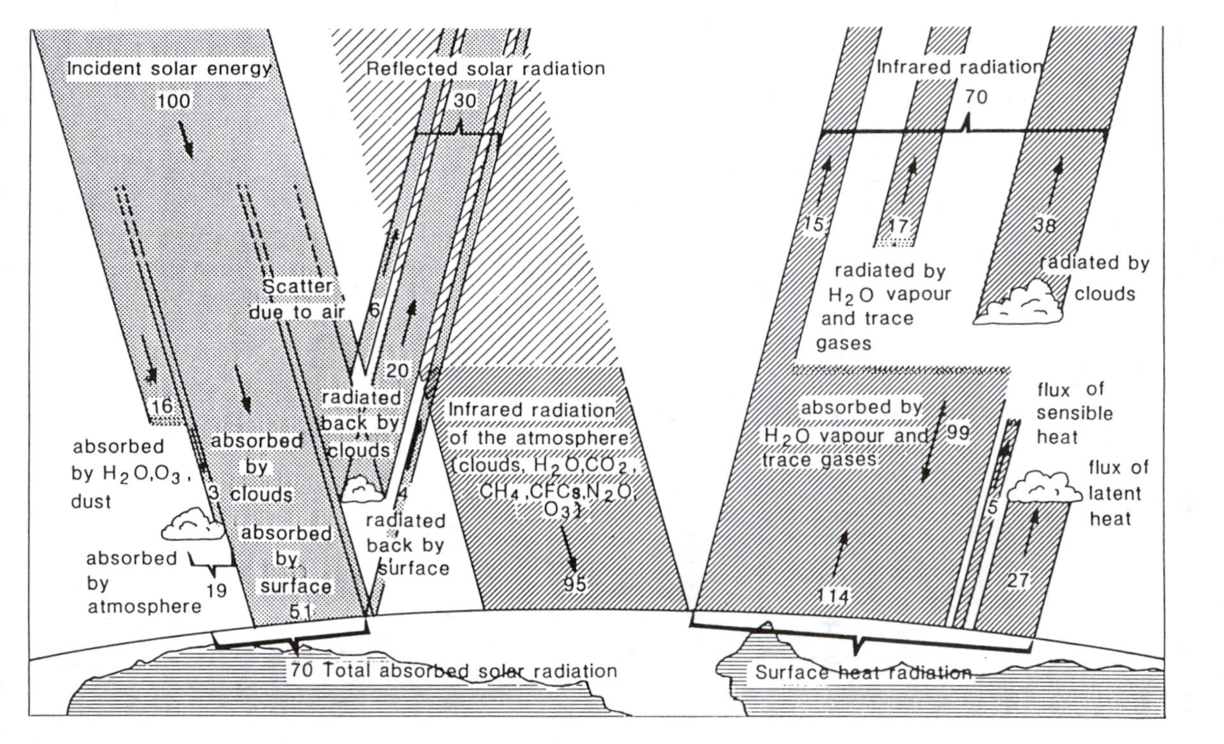

Fig. 5.2. Radiation budget of the earth-atmosphere system and distribution of incidence of solar radiation at the surface level and in the atmosphere. Totals are shown at the boundaries of the atmosphere and the surface of the earth. Short-wave radiation fluxes are shown on the left, and long-wave radiation fluxes of the earth and the atmosphere are shown on the right. In both cases, the incidence of solar radiation at the outer boundary of the atmosphere is used as a reference (100). Changes in the concentrations of the greenhouse gases promote changes in temperature. (Adapted from German Bundestag, 1991)

Table 5.1. Atmospheric trace gases that are radiatively active and of significance to global change[a]

Characteristic	**Carbon dioxide (CO_2)**	**Carbon monoxide (CO)**	**Chlorofluoro-carbons (CFCs)**	**Methane (CH_4)**	**Nitrous oxide (N_2O)**	**Tropospheric ozone (O_3)**	**Water vapor (H_2O)[b]**
Principal anthropogenic sources	Fossil fuels; deforestation	Fossil fuels; biomass burning	Refrigerants; aerosols; industrial processes	Rice culture; cattle; fossil fuels; biomass burning	Fertilizer; land use conversion	Hydrocarbons (with NO_x); biomass burning	Land conversion; irrigation
Principal natural sources	Balanced in nature	Hydrocarbon oxidation	None	Wetlands	Soils; tropical forests	Hydrocarbons	Evapotranspiration
Atmospheric lifetime	50–200 years	Months	60–100 years	10 years	150 years	Weeks to months	Days
Present atmospheric concentration **(ppb at surface)**	353,000	100[c]	CFC-11: 0.28 CFC-12: 0.48	1,720	310	25–45[c]	3,000–6,000 in stratosphere
Preindustrial concentration (1750–1800) **(ppb at surface)**	280,000	40–80	0	790	288	10	Unknown
Present annual rate of increase (%)	0.5	0.7–1.0[c]	4	0.9	0.3	0.5–2.0[c]	Unknown
Relative contribution to the anthropogenic greenhouse effect (%)	60	0	12	15	5	8	Unknown

[a] Adapted from EarthQuest, 1990.
[b] Can contribute to warming or cooling, depending on whether it is in the low clouds (cooling) or high clouds (warming).
[c] Northern hemisphere.

example, with overhead sun, a 10% decrease in typical O_3 amounts was predicted to result in a 20% increase in UV-B penetration at 305 nm, a 250% increase at 290 nm, and a 500% increase at 287 nm, all wavelengths within the UV-B band (Cutchis, 1974).

With or without these predicted changes, the incoming solar radiation to the earth's surface is of short wavelength (see Table 1.2). After some absorption, surfaces radiate heat energy back to the atmosphere at longer wavelengths (in the infrared region). This energy is trapped by certain atmospheric chemical constituents and by clouds, leading to a warming of the atmosphere above the earth's surface (Fig. 5.2 and Table 5.2). This process is known as the natural greenhouse effect, without which the planet would not be habitable. The critical concern today is whether human influence has increased and accelerated this greenhouse effect toward progressive global warming, which ultimately could lead to disastrous ecological consequences (Houghton and Woodwell, 1989).

Surface emissions and concentrations of globally important trace gases are increasing (Table 5.1). Many of these gases can have direct effects on the climate by trapping infrared radiation. Climate modification, associated with long-term changes in weather, is characterized by concerns regarding trends and variability in surface temperatures, precipitation patterns, cloud cover, and other climatic variables. The absorption of surface-emitted, outgoing infrared radiation in the atmosphere, followed by reemission at the local atmospheric temperature, can lead to an increase in surface temperature, known as the modified greenhouse effect (Table 5.3).

As previously stated, climate modification and the greenhouse effect are governed by interactions between stratospheric and tropospheric processes. According to McElroy and Salawitch (1989), a panel of experts convened by NASA (the U.S. National Aeronautics and Space Administration) concluded that the best current analysis of data, mainly

Table 5.2. Contribution of trace gases to the natural greenhouse effect of 33°C[a]

	Contribution[b]	
Trace gas	**°C**	**Percentage**
Carbon dioxide	7.2	22
Methane	0.8	2.4
Nitrous oxide	1.4	4.2
Ozone	2.5	7.3
Water vapor	20.6	62

[a] Adapted from German Bundestag, 1991.
[b] The balance is accounted for by a number of other trace gases.

from ground-based measurements, indicates that the annual average column density of O_3 declined 1.7–3.0% in the latitude band 30–64°N between 1969 and 1986. The period covered by this analysis occupies less than one solar cycle and includes two significant geophysical events, the eruption of the volcano El Chichon in Mexico and the unusually large El Niño (the flow of warm ocean waters from the southern to the northern hemisphere that results in oscillations in air temperature). There have also been problems in the past with satellite-based instrumentation. Nevertheless, according to NASA, model calculations are broadly consistent with the observed changes in column O_3, except that the mean values of observed decreases at mid- and high latitudes during the winter are larger than the predicted values.

A consequence of the measured or predicted stratospheric O_3 depletion is the increased penetration of radiation in the UV-B band (280 to 320 nm) into the lower troposphere. According to Frederick et al. (1989), biologically effective UV-B irradiation at the earth's surface varies with 1) the elevation of the sun, 2) the amount of atmospheric O_3, and 3) the abundance of atmospheric matter generated by natural and human-made processes that have scattering and absorbing properties. Taken alone, the reported decrease in the O_3 column over the northern hemisphere between 1969 and 1986 implies an increase in erythemal (**erythema** is a redness of the skin, such as that caused by sunburn) irradiation at the ground of $\leq 4\%$ during the summer. However, an increase in tropospheric absorption caused by polluting gases or particulate matter over localized areas could more than offset the predicted enhancement in UV-B or global radiation. Any such extra absorption is

Table 5.3. Some predictions of global average equilibrium response (climate sensitivity) of the surface temperature to a doubling of the ambient CO_2 concentration[a]

Temperature increase (°C)	Source
0.5–1.2	Lindzen et al. (1982) Ramanathan (1981) IPCC (1990)
1.0–5.0	National Academy of Sciences (1992)
1.4–3.0	Schlesinger et al. (1992)
1.5–4.5 (best estimate 2.5)	IPCC (1992a,b)
1.0–3.5	IPCC (1995)
2.0–4.0	Hansen et al. (1991) Hansen et al. (1993)

[a] Adapted from Karl, 1993. Refer to that paper for the other references listed in this table. IPCC (1995) is an exception.

likely to be highly regional in nature and does not imply that a decrease in erythemal radiation has occurred on a global basis.

Most recently, Kerr and McElroy (1993) found that spectral measurements of UV-B radiation taken in Toronto, Canada, since 1989 indicate that the intensity of light at wavelengths near 300 nm has increased by 35% per year during the winter (low levels of surface O_3 and fine particles) and 7% per year during the summer (high levels of surface O_3 and fine particles). The authors attribute these results to a downward trend in the total (stratosphere + troposphere) O_3 column. Frederick et al. (1991) also reported an increase in surface-level, biologically effective radiation of 0.62–7.56% per decade at several geographic locations (e.g., Bismarck, North Dakota, and Boulder, Colorado, in the United States; Belsk, in the former U.S.S.R.; and Uccle, Belgium, and Arosa, Switzerland, in Europe). In this context, the observed and much publicized Antarctic **O_3 hole** may represent a major occurrence, where a portion of the earth has experienced UV-B radiation levels during the spring that are far in excess of levels that prevailed prior to the present decade.

A conclusion that can be derived from the study by Frederick et al. (1989) and the numerous studies of spatial variability of air pollutants and their deposition patterns is that the use of average values across geographic areas is inappropriate, since spatial variability or geographic patchiness are not considered. Nevertheless, changes in global surface temperature have been estimated to be +0.7°C during the past 140 years and between +1.5 and +4.5°C from the 19th to the 21st centuries (Wuebbles et al., 1989). This increase in temperature is considered to be the result of the increased trapping of infrared reradiation from the earth's surface by the increasing concentrations of tropospheric gases, for example, CO_2 (Table 5.1). Different tropospheric gases vary in their characteristics relative to warming. For example, methane (CH_4), although present at concentrations approximately 200 times lower than those of CO_2 (Table 5.1), is considered to be 15–30 times more effective than CO_2.

Present-day tropospheric CO_2 concentrations are predicted to double (600–700 ppm) during the 21st century; CH_4 concentrations are currently increasing at an annual rate of about 0.9% and N_2O levels by about 0.3% (Table 5.1). Nevertheless, problems arise with the use of average values in concluding that global air temperature appears to have increased by roughly 0.7°C during the past 140 years (NASA, 1988). The problems include 1) uncertainties associated with the historical database of air temperatures over oceans, where measurement methods have improved with time and the correction factors are in question, and 2) the location of many land-based measurement devices in or close to urban centers (heat islands) rather than in rural settings. These types of uncertainties have resulted in controversy concerning global warming. Of equal concern is the need to separate the effects of natural geophysical and chemical cycles from any observed and/or perceived changes in the global climate resulting from human influences.

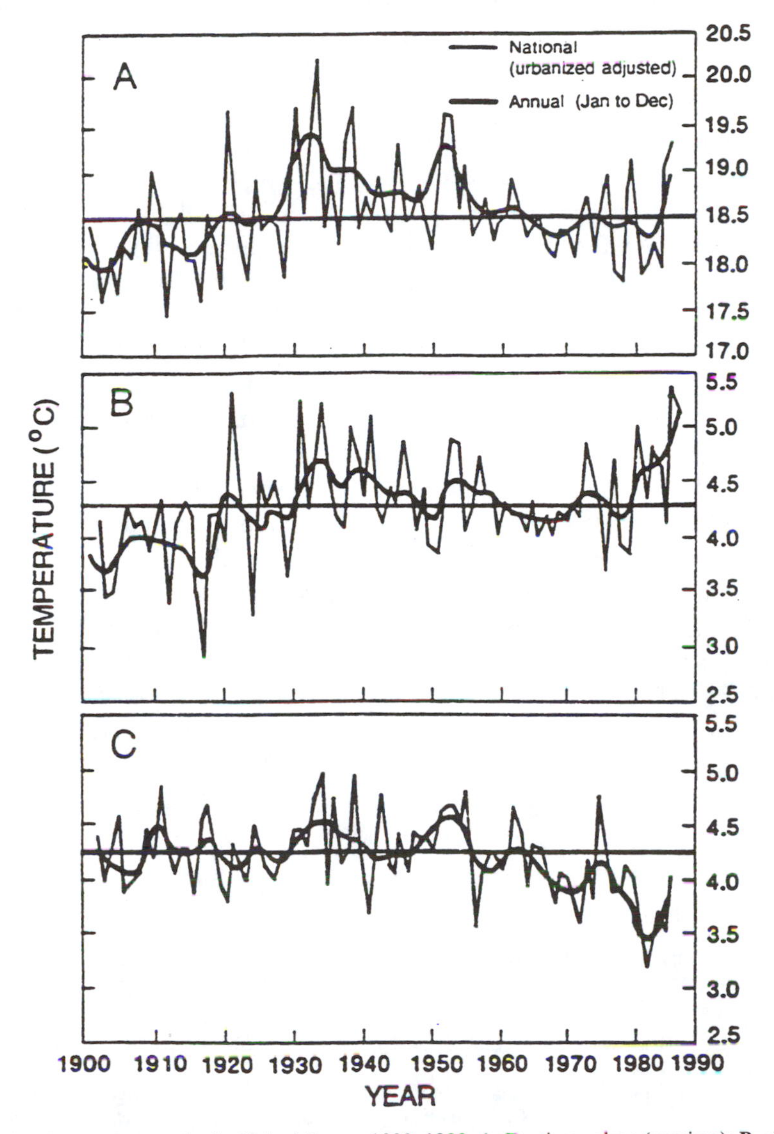

Fig. 5.3. Air temperatures in the United States, 1900–1990. **A,** Daytime values (maxima); **B,** nighttime values (minima); and **C,** differences between daytime (maxima) and nighttime (minima) values. (Adapted from Karl et al., 1988)

Karl et al. (1988) analyzed temperature data from 492 deurbanized, historical climate network stations in the United States. According to their results, the difference between day and night temperatures has declined since about 1950 (Fig. 5.3). During this period, daytime temperatures have declined slightly and nighttime temperatures have risen. Similar results for day and night temperatures were also obtained in mainland China and the former Soviet Union (Karl et al., 1991). Although the exact cause of these changes is not fully understood, increase in cloud cover and atmospheric concentrations of sulfate aerosols most likely were major determinants.

In contrast to the study by Karl et al. (1988), any predictions of future global climate change based on one- or two- rather than three-dimensional circulation models are disturbing. Certainly the application of three-dimensional models is limited by the present-day availability of computing power, although this is changing rapidly. Nevertheless, global change predictions are based on **general circulation models (GCMs)** of global scale. Of additional concern is the fact that many of these models have not considered the role of clouds until noon. Recent studies on the role of clouds, based on the Earth Radiation Budget Experiment (ERB), show atmospheric cooling over North America (Ramanathan et al., 1989). Clouds appear to have a net cooling effect globally equal to about four times as much energy as would be trapped by doubling the CO_2 levels. In mid- and high latitudes, the net cooling effect of clouds is large, but over the tropics, cooling is nearly equalized by heating.

Whether or not global warming is occurring, in the context of terrestrial vegetation, atmospheric **processes** are as important as their **products** (Table 5.4). Some scientists tend to overlook this important consideration in their preoccupation with assessing the impacts of global warming (Parry, 1990).

Table 5.4. Summary of atmospheric processes and products relevant to global climate change[a]

Process	Products
Loss of stratospheric O_3	Increased UV-B penetration[b]
Increases in greenhouse gases:[c] CO_2, O_3, CO, CH_4, CFCs, OBs, N_2O, COS, SO_2?, NO_x and ($NO + NO_2$); and changes in the concentration of H_2O molecules	Changes in temperature, precipitation, radiation, evaporation, wind, and secondary aerosols
Increased UV-B penetration[b]	Decreased O_3

[a] Source: Runeckles and Krupa, 1994.
[b] Increased UV-B penetration may decrease tropospheric O_3 and affect other greenhouse gases.
[c] CFCs = chlorofluorocarbons; CH_4 = methane; CO = carbon monoxide; CO_2 = carbon dioxide; COS = carbonyl sulfide; H_2O = water; N_2O = nitrous oxide; NO = nitric oxide; NO_2 = nitrogen dioxide; NO_x = oxides of nitrogen; O_3 = ozone; OBs = organo-bromines; SO_2 = sulfur dioxide; and UV-B = radiation at 280 to 320 nm.

A Brief Assessment

During the next 50 years, if the word "climate" is considered in its broadest sense, we will see changes in the atmospheric concentrations of carbon dioxide (CO_2), patterns of temperature and moisture, levels of UV-B (280 to 320 nm) radiation reaching the earth's surface, amounts and distribution of tropospheric ozone (O_3) and fine particulate matter, and probably the frequency of high winds (Lodge, 1990). Some of these changes will be related to "natural" climatic shifts, while others will be directly related to human activities. Most researchers believe that the climate will simply become warmer worldwide (Parry, 1990). While this may be true on a global mean basis, this is by no means necessarily true for a given spot on the earth or even for a given nation or continent. Instead, sizable areas may well become warmer, cooler, drier, or wetter or remain unchanged, insofar as annual means are concerned. Probably any modification of the climate will manifest itself through changes not in mean values, but in the deviations from those means and in the frequency of severe weather conditions such as high winds, thunderstorms, and blizzards (Lodge, 1990).

Whether surface air temperatures will increase and, if so, by how much, where, and when will continue to be controversial issues. Increases in the ambient concentrations of trace gases such as CO_2 and O_3 (Table 5.1) by themselves will continue to have an effect on vegetation (Krupa and Kickert, 1989). This subject is discussed further in Chapter 7, Air Quality and Crops, and Chapter 8, Air Quality and Forests.

References

Cicerone, R. J. 1987. Changes in stratospheric ozone. Science 237:35-42.

Cutchis, P. 1974. Stratospheric ozone depletion and solar ultraviolet radiation on earth. Science 184:13-19.

EarthQuest. 1990. University Corporation for Atmospheric Research, Boulder, CO.

Frederick, J. E., Snell, H. E., and Haywood, E. K. 1989. Solar ultraviolet radiation at the earth's surface. Photochem. Photobiol. 50:443-450.

Frederick, J. E., Weatherhead, E. C., and Haywood, E. K. 1991. Long-term variations in ultraviolet sunlight reaching the biosphere—Calculations for the past 3 decades. Photochem. Photobiol. 54:781-788.

German Bundestag. 1991. Protecting the Earth. A Status Report with Recommendations for a New Energy Policy. Deutscher Bundestag, Referat Öffentlichkeitsarbeit, Bonn.

Houghton, R. A., and Woodwell, G. M. 1989. Global climatic change. Sci. Am. 4:36-44.

Intergovernmental Panel on Climate Change (IPCC). 1995. IPCC Working Group I 1995 Summary for Policymakers. Proc. WGI Sess., 5th. IPCC WGI Technical Support Unit.

Johnston, H. S. 1971. Reduction of stratospheric ozone by nitrogen oxide catalysts from supersonic transport exhaust. Science 173:517-522.

Karl, T. R. 1993. Missing pieces of the puzzle. Res. Explor. 9:235-249.

Karl, T. R., Baldwin, R. G., and Burgin, M. G. 1988. Historical climatology, ser. 4 and 5. National Climatic Data Center, Asheville, NC.

Karl, T. R., Kukla, G., Razuavayev, V. N., Vyacheslav, N., Changery, M., Quayle, R. G., Heim, R. R., Easterling, D. R., and Cong, B. F. 1991. Global warming: Evidence for symmetric diurnal temperature change. Geophys. Res. Lett. 18:2252-2256.

Kerr, J. B., and McElroy, C. T. 1993. Evidence for large upward trends of ultraviolet-B radiation linked to ozone depletion. Science 262:1032-1034.

Krupa, S. V., and Kickert, R. N. 1989. The greenhouse effect: Impacts of ultraviolet-B (UV-B) radiation, carbon dioxide (CO_2), and ozone (O_3) on vegetation. Environ. Pollut. 61:263-393.

Krupa, S. V., and Manning, W. J. 1988. Atmospheric ozone: Formation and effects on vegetation. Environ. Pollut. 50:101-137.

Lodge, J. P., Jr. 1990. Climate change in the context of forest response. Paper 90-152.4 in: Proc. Annu. Meet. Air Waste Manage. Assoc., 83rd. Air Waste Management Association, Pittsburgh, PA.

McElroy, M. B., and Salawitch, R. J. 1989. Changing composition of the global stratosphere. Science 243:763-770.

Molina, M. J., and Rowland, F. S. 1974. Stratospheric sink for chlorofluoromethanes: Chlorine atom catalyzed destruction of ozone. Nature 249:810-812.

National Aeronautics and Space Administration (NASA). 1988. Earth System Science, A Closer View. Report of the Earth System Sciences Committee. NASA Advisory Council, Washington, DC.

Parry, M. L. 1990. Climate Change and World Agriculture. Earthscan, in association with the International Institute for Applied Systems Analysis and United Nations Environment Programme, London.

Ramanathan, V., Cess, R. D., Harrison, E. F., Minnis, P., Barkstrom, B. R., Ahmad, E., and Hartmann, D. 1989. Cloud-radiative forcing and climate: Results from the Earth Radiation Budget Experiment. Science 243:57-63.

Runeckles, V. C., and Krupa, S. V. 1994. The impact of UV-B radiation and ozone on terrestrial vegetation. Environ. Pollut. 83:191-213.

Wuebbles, D. J., Grant, K. E., Connell, P. S., and Penner, J. E. 1989. The role of atmospheric chemistry in climate change. J. Air Pollut. Control Assoc. 39:22-28.

Further Reading

Frederick, K. D., and Rosenberg, N. J. 1994. Assessing the Impacts of Climate Change on Natural Resource Systems. Kluwer Academic Publishers, Dordrecht, Netherlands.

Gates, D. M. 1993. Climate Change and Its Biological Consequences. Sinauer Associates, Sunderland, MA.

Graedel, T. E., and Crutzen, P. J. 1995. Atmosphere, Climate and Change. W. H. Freeman, New York.

Intergovernmental Panel on Climate Change (IPCC). 1990. Climate Change: The IPCC Scientific Assessment. J. T. Houghton, G. J. Jenkins, and J. J. Ephraums, eds. Cambridge University Press, Cambridge.

Rowland, F. S. 1989. Chlorofluorocarbons and the depletion of stratospheric ozone. Am. Sci. 77:36-45.

CHAPTER 6

Air Quality and Human Health

> If the air is altered ever so slightly, the state of the Psychic Spirit will be altered perceptibly. Therefore, you find many persons in whom you can notice defects in the actions of the psyche with the spoilage of air, namely, that they develop dullness of understanding, failure of intelligence and defect of memory . . .
>
> —Maimonides; quoted by V. Goodhill

Introduction

Deterioration of air quality and alterations in human disposition have been known since the Roman Empire (Seneca, 61 A.D.; in Stern et al., 1973). During the 20th century, deaths and illnesses have been related to major air pollution episodes, for example, in Meuse Valley, Belgium (December, 1930); Donora, Pennsylvania (October, 1948); London, England (December, 1952); and Bhopal, India (December, 1984). In all these cases, human populations were subjected to very high concentrations of various toxic air pollutants (e.g., SO_2 and particulate matter or methyl isothiocyanate) from a few days to weeks. These types of exposures are known as **acute exposures,** and generally the corresponding responses are acute, progressively declining over a prolonged time. Fortunately, disasters such as that in Bhopal are rare. In contrast, in the present-day developed and developing countries, particularly in large urban centers and in the vicinities of major emission sources, human populations and other environmental receptors are subjected continuously to relatively low-level concentrations of various air pollutants with intermittent, periodic, random episodes of relatively high concentrations lasting from a few minutes to 1 day or longer. These types of exposures are known as **chronic exposures,** and the corresponding responses are also chronic in nature. Given the heterogeneity (diversity of genetic back-

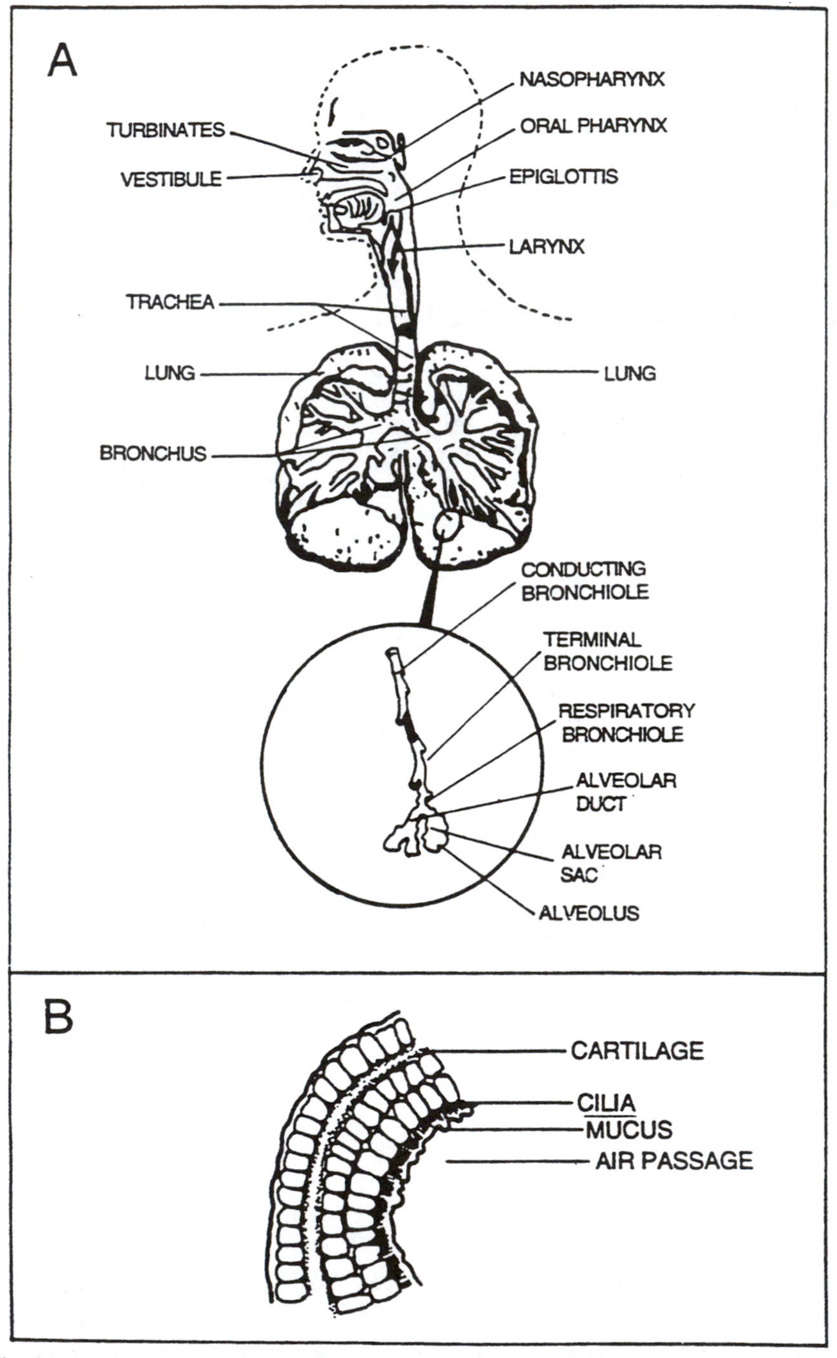

Fig. 6.1. A, Human respiratory tract and **B,** internal structure of respiratory airways. (Adapted from Hinds, 1982, and Godish, 1991)

ground, living habits, working environment, and mobility) of human populations and the frequent impracticality of using humans as subjects in exposure experiments, direct proof of chronic air pollutant exposure and clinical response is hard to come by. Frequently, such relationships have to be derived from **epidemiological** (the sum of factors controlling the presence or absence of a particular disease) and some animal-exposure studies. Nevertheless, air pollution is considered to be a direct contributor to accelerated aging, **asthma**, **berylliosis**, **emphysema**, and **mesothelioma** and a contributing cause of **bronchitis** and cancer of the gastrointestinal and respiratory tracts (Table 6.1 and Fig. 6.1). See Notes 6.1 for definitions of these terms.

Health Effects of Some Air Pollutants

Carbon Monoxide (CO)

Carbon monoxide is a primary pollutant. The immediate effect of CO is that it chemically binds with hemoglobin in the blood to form **carboxyhemoglobin (COHb)**. Carbon monoxide's affinity for hemoglobin is 200 times greater than that of oxygen (O_2), and it also tends to remain more tightly bound. Carbon monoxide is highly toxic at concentrations >1,000 ppm, leading to death from asphyxiation because the body, particularly the brain, is deprived of sufficient O_2.

Carbon monoxide is colorless, odorless, and tasteless, and therefore humans are unaware that they are being exposed to it. Furthermore, CO tends to diffuse from near the surface in enclosed spaces, such as a closed garage. This is particularly important for houses with attached garages, since many individuals tend to allow car engines to run in their

Table 6.1. Diseases for which evidence points to air pollutants as either direct or contributing causes[a]

Disease	Air pollutants involved
Accelerated aging	Ozone and oxidant air pollutants (direct)
Allergic asthma	Airborne denatured grain protein, etc. (direct)
Berylliosis	Airborne beryllium compounds (direct)
Bronchitis	Acid gases; particulates; respiratory infection; inclement weather (contributing)
Cancer	
Gastrointestinal tract	Airborne carcinogens; hereditary tendency (contributing)
Respiratory tract	Airborne carcinogens; hereditary tendency (contributing)
Emphysema	Airborne respiratory irritants and familial tendency (direct)
Mesotheliomas	Asbestos, associated trace metals, carcinogens in air, other fibers (?) (direct)

[a] Adapted from Stern et al., 1973.

garages during cold weather. This is dangerous, and cars should be moved outside the garage to run and warm up.

Death by CO poisoning is not uncommon, and it is frequently the cause of death in structural fires, including home fires (CO asphyxiation). The amount of COHb produced in the blood is directly related to CO dose (concentration and duration of exposure). Exposures to 35 ppm CO for 1 hr or 9 ppm for 8 hr will result in 1.3% COHb saturation. These two exposure values represent the **ambient air quality standard** for CO in the United States in 1995 (see Table 2.11) to protect human health. Table 6.2 shows the relationship between COHb levels and health effects, resulting mostly from the brain's response to reduced O_2

Notes 6.1. Definitions of some technical terms used in clinical medicine

Asthma	= A condition often marked by labored breathing accompanied by wheezing.
Berylliosis	= Multiple nodule–like accumulation of inflamed tissue in the lung, the nodules exhibit granulation; also enlargement of the lymph nodes.
Bronchitis	= Inflammation of the bronchial tubes.
Emphysema	= A condition of the lung marked by enlargement and frequently by impairment of heart action.
Mesothelioma	= A tumor derived from the mesothelial tissue such as the membrane folded over the surface of the lung.

Table 6.2. Human responses to short-term CO exposures[a]

Blood COHb[b] levels (%)	Effect
0–1	None
2.5	In nonsmokers, impairment of time-interval discrimination
2.8	In exercising patients, time to onset of angina pectoris pain shortened and duration of pain lengthened[c]
3	Changes in relative brightness thresholds
4.5	Increased reaction time to visual stimuli
10	Changes in performance in driving simulation
10–20	Headache, fatigue, dizziness, loss of coordination

[a] Adapted from Godish, 1991.
[b] Carboxyhemoglobin.
[c] Angina pectoris is a disease characterized by brief, sudden attacks of chest pain precipitated by deficient oxygenation of the heart muscle.

levels. In addition to these effects related to the central nervous system, exposures to CO can contribute to or increase the severity of cardiovascular disease, such as the hardening of arteries. This can result in the interruption of the blood supply to the heart muscle, chest pain, and sudden death. Evidence suggests that CO-induced hardening of the arteries is associated with high-fat, high-cholesterol diets. However, there is no evidence to indicate a relationship between ambient CO exposures and the incidence of heart attacks, excluding individuals frequently exposed to smoke-laden environments.

Ozone (O_3)

Ozone is one of the most toxic secondary air pollutants in the context of human health. It can induce measurable physiological changes in humans (Table 6.3) at concentrations and exposures that are within the range of polluted ambient environments (National Research Council, 1991).

A significant amount of research has been conducted under controlled laboratory conditions on the effects of O_3 on animals and human volunteers. These studies have demonstrated significant changes in lung function in response to exposures to 0.10–0.40 ppm O_3 for 1–2 hr. The 1995 **ambient air quality standard** for O_3 in the United States is 0.12 ppm for 1 hr, not to be exceeded more than once a year at a given location (see Table 2.11). The observed O_3 effects in laboratory studies include increased respiratory rate, increased **pulmonary** (relating to the lungs) resistance, decreased **tidal volume** (rise and fall) of air intake, and changes in respiratory mechanics. These changes, however, are transient. The issue of whether smokers are more susceptible to these O_3 effects is a controversial one, because negative effects have been reported for healthy adolescent individuals and young adults (Godish, 1991). Heavily exercising adults report cough, shortness of breath, and pain on deep air intake in response to O_3 exposures of 0.12 ppm. Above 0.12 ppm, they report symptoms such as throat dryness, chest tightness, pain within the chest, coughing, wheezing, pain on deep air intake, shortness of breath, weariness, indefinite feeling of debility, headache, and nausea. It is important to note here that copying machines and other instruments that utilize high-voltage current (corona discharge) can be sources of O_3 pollution indoors on days with low humidity. Therefore, such equipment should not be placed in areas where people work all day. It is also important to state that although many of the previously described symptoms have been observed in adults, such effects apparently do not occur in children exposed to the same levels of O_3.

Table 6.3. Gradation of individual physiological responses to acute O_3 exposure[a]

Response	Mild	Moderate	Severe	Incapacitating
Change in spirometry $FEV_{1.0}$ and FVC[b]	5–10%	10–20%	20–40%	>40%
Duration of effect	Complete recovery in <30 min	Complete recovery in <6 hr	Complete recovery in 24 hr	Recovery in >24 hr
Symptoms	Mild to moderate cough	Mild to moderate cough, pain on deep inspiration, shortness of breath	Repeated cough, moderate to severe pain on deep inspiration and shortness of breath; breathing distress	Severe cough, pain on deep inspiration and shortness of breath; obvious distress
Limitation of activity	None	Few individuals choose to discontinue activity	Some individuals choose to discontinue activity	Many individuals choose to discontinue activity

[a] Reprinted from U.S. Environmental Protection Agency, 1988.

[b] **Spirometry** = the measurement of the volume of air entering and leaving the lungs. **$FEV_{1.0}$** = forced expiratory (release of air from the lungs) volume in the **first** second of a vital capacity maneuver; and **FVC** = forced vital capacity (the amount of air that can be forcibly expelled from the lungs following a full inspiration).

Overall, ambient O_3 exposures at certain geographic locations can reduce athletic performance. In addition, O_3 can interfere with or inhibit the ability of the immune system to defend the body against microbial infections. Ozone-induced changes in pulmonary defense mechanisms in animals include 1) impaired ability to inactivate bacteria; 2) impaired **macrophage phagocytic activity** (activity of a large cell that engulfs foreign matter); and 3) reduced lysosomal enzymatic activity (a **lysosome** is a saclike organelle that contains various hydrolytic enzymes).

While the ambient air quality standard is based on a 1-hr peak exposure (0.12 ppm; see Table 2.11), ambient O_3 episodes at or near 0.12 ppm can last for several hours in a row on one or more days. Such prolonged exposures can result in progressive changes in respiratory function, a marked increase in nonspecific airway reactivity, and progressive changes in symptoms. Repetitive exposures result in an enhanced response on the second day, decreasing responses on the third and fourth days, and virtually no response on the fifth day. This apparent adaptation persists for a week after the exposure. Epidemiological studies of chronic human exposure to ambient air suggest that there may be a functional adaptation that persists for several months after the high O_3 season and that the body returns to normal during the following spring. According to Lippmann (1989), the effects of long-term chronic exposure to O_3 remain poorly defined, but epidemiological and animal inhalation studies suggest that current ambient levels are sufficient to cause premature aging of the lungs.

Oxides of Nitrogen (NO_x)

Nitric oxide (NO) at current ambient concentrations is not considered a threat to human health; however, it is rapidly converted to NO_2 (nitrogen dioxide) in the atmosphere. In contrast to SO_2, which is rapidly absorbed by the fluids in the upper tracheobronchial zone (Fig. 6.1), NO_2 is less soluble and therefore can deeply penetrate the lungs, leading to tissue damage. Adverse effects such as pulmonary **edema** (an abnormal, excess accumulation of watery fluids in the connective tissue) usually do not appear until many hours after exposure to high concentrations. Such exposures are known to occur during the manufacture of nitric acid (HNO_3), during electric arc welding, and in farm silos.

Animal toxicological studies show that exposures to $\geq$0.50 ppm NO_2 can result in destruction of cilia, disruption of alveolar tissue, and obstruction of respiratory bronchioles (Fig. 6.1). Such studies also show that NO_2 may cause or aggravate respiratory infections.

Hydrocarbons (HCs)

The hydrocarbons and their derivatives that are of health concern include the carcinogenic polyaromatic hydrocarbons (PAHs) such as benzo[*a*]pyrene (BaP) (see Fig. 2.5) and eye irritants such as formaldehyde (HCHO) and acrolein. Benzo[*a*]pyrene is the most abundant PAH in urban air. Controlled animal studies show that BaP exposures far in excess of the ambient levels are needed to induce cancer.

In evaluating the carcinogenic potential of PAHs, microbial mutagenicity (ability to cause a mutation, e.g., in the bacterium *Salmonella*) has been used as a reasonably good predictor of risk. In addition to *Salmonella*, cultured mammalian cells and human lymphocytes have also been used in the testing. Table 6.4 provides a list of some of the known carcinogenic PAHs.

Sulfur Dioxide (SO_2)

Because of its solubility in water and related fluids, inhaled SO_2 is almost entirely removed in the mouth, throat, and nose; therefore, <1% of the SO_2 reaches the lung tissue. The principal effect of SO_2 is the alteration of the mechanical function of the upper airway. This includes an increase in nasal flow resistance and a decrease in nasal mucus flow rate. Exposing strenuously exercising asthmatic subjects to relatively low levels of SO_2 (0.25 and 0.50 ppm) produces acute bronchial constriction on inhalation. This hypersensitive response can also be caused by the use of sodium bisulfite as a food preservative (on salad greens) in restaurants and in the production of wine, since SO_2 is produced upon the decomposition of sodium bisulfite. This has led the U.S. Food and Drug Administration (FDA) to apply restrictions on the use of sodium

Table 6.4. Acronyms, molecular weights, and carcinogenic activities of some environmentally relevant polyaromatic hydrocarbons (PAHs)[a]

PAH	Acronym	Molecular weight	Carcinogenic activity[b]
Benz[*a*]anthracene	BaA	228.0939	+
Benzo[*a*]pyrene	BaP	252.0939	++
Benzo[*e*]pyrene	BeP	252.0939	0/+
Benzo[*ghi*]perylene	BghiP	276.0939	+
Chrysene	CHR	228.0939	0/+
Coronene	COR	300.0939	0/+
Cyclopenta(*cd*)pyrene	CPY	226.0783	+
Fluoranthene	FL	202.0783	+
Quinoline	. . .	129.0578	+

[a] Adapted from Finlayson-Pitts and Pitts, 1986.
[b] + = carcinogenic; ++ = highly carcinogenic; and 0/+ = a range of results from no effect to carcinogenic.

bisulfite in restaurants. Some of the health effects attributed to SO_2 most likely result from its conversion to fine-particle sulfate aerosols such as H_2SO_4 (sulfuric acid).

Particulate Matter (PM)

The health effects of atmospheric particulate matter depend on its ability to penetrate respiratory defense mechanisms. In the nasal region, for example, large particles may be removed by stiff nasal hairs or by impingement on the mucus layer. Cilia sweep the mucus layer and the entrapped particles toward the back of the mouth where they may be swallowed or ejected by spitting. Foreign particulate matter may also be removed by sneezing (the average velocity of a sneeze is ~200 km per hour). The irritating effect of foreign particulate matter initiates a respiratory muscle contraction by closing the vocal cords in order to build up high pressure. Suddenly, the vocal cords open, expelling pressurized air and any foreign matter in its path.

In addition to these responses, when foreign matter enters the upper portion of the bronchial tree, muscle layers constrict the bronchi, narrowing the bronchial diameter and thus reducing the amount of material that can enter the lungs. The cough reflex, which is similar to the sneeze reflex, assists in this process by eliminating mucus that has trapped particles or removed them from deeper lung tissue. The cilia that line the tracheobronchial system also clear the airways by propelling mucus and foreign matter upward, where it is removed by coughing, spitting, or swallowing (the entrance to gastrointestinal system).

In general, respiratory defense mechanisms are sufficient to remove particles >10 μm in diameter from inhaled air. Therefore, in the United States the **National Ambient Air Quality Standard** (see Table 2.11) is for **PM-10** (particles in the size range of <10 μm). Particles <10 μm or **inhalable** particles can enter and be deposited in the respiratory system. Particles <2.5 μm are called **respirable**, since they can enter and be deposited in the pulmonary system. In contrast, particles >2.5 μm are removed in the upper respiratory system (Fig. 6.2).

Because of changes in airflow patterns in the tracheobronchial zone, particles tend to be deposited at or near airway bifurcations. Since nerve endings are concentrated at these sites, the mechanical stimulus from the deposited particles often leads to reflex coughing and bronchial constriction. The sensitivity of nerve endings to chemical stimuli results in an increase in breathing rate and reduced **pulmonary compliance** (ability of the lung to yield to increases in pressure without disruption).

The deposition of particles is influenced not only by their size but also by their concentration, molecular or chemical composition, pH, and solubility. Deposition also varies among nonsmokers, smokers, and individuals with lung disease. Tracheobronchial deposition is slightly higher in smokers and greatly increased in individuals with lung disease.

After particles have been deposited, their retention may be a function of the rate of clearance, which varies greatly among different regions of

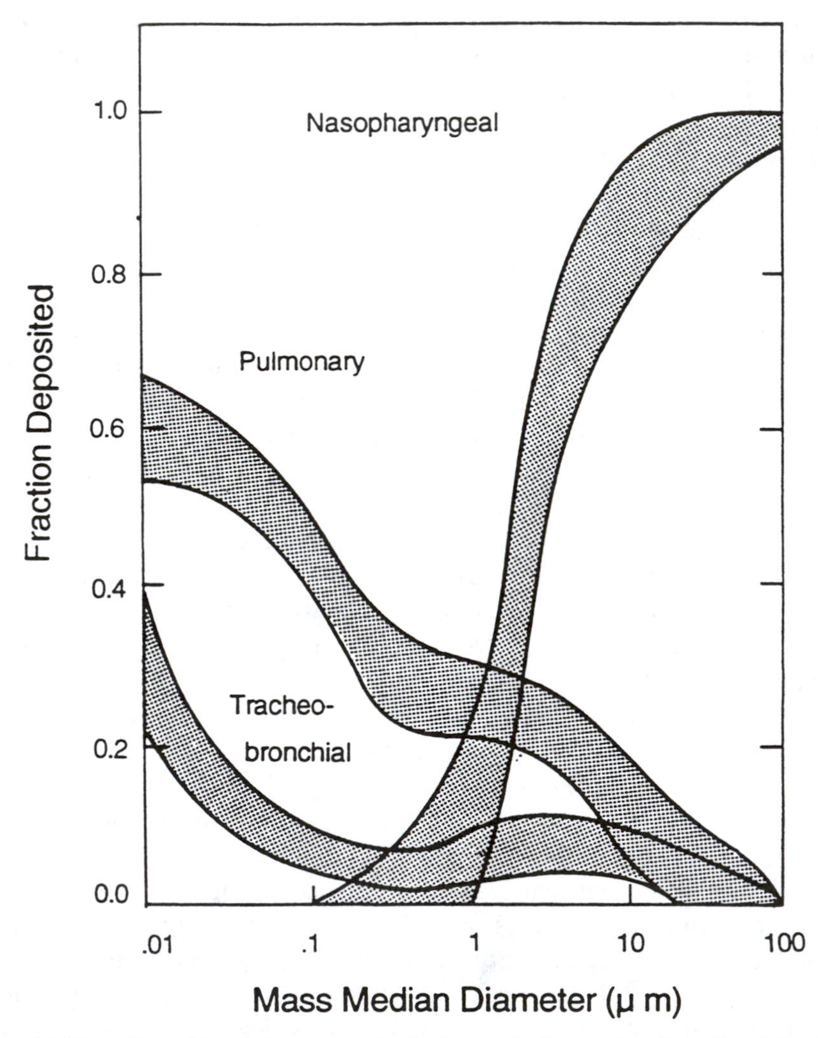

Fig. 6.2. Deposition of particulate matter in the human body as a function of particle size. (Reprinted from U.S. Department of Health, Education and Welfare, 1969)

1.1–1.4. Geographic and temporal relationships between visibility (governed by the accumulation of fine particles in the atmosphere), ambient ozone (O_3), and path of long-range transport of air masses. Note the similarity of the geographic patterns in the reduction of visibility and the occurrence of O_3. (Reprinted from Krupa et al., 1982)

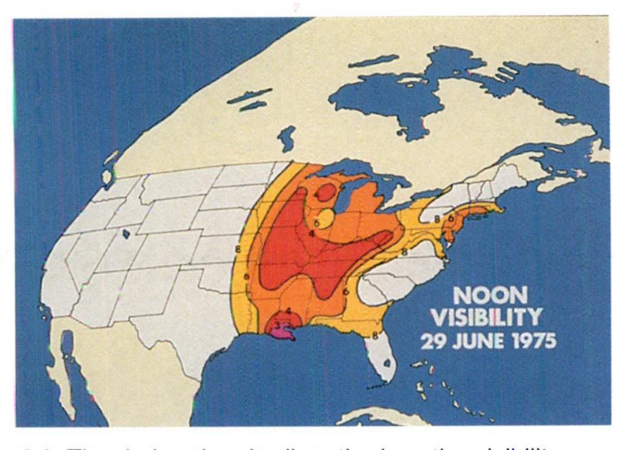

1.1. The darker the shading, the less the visibility.

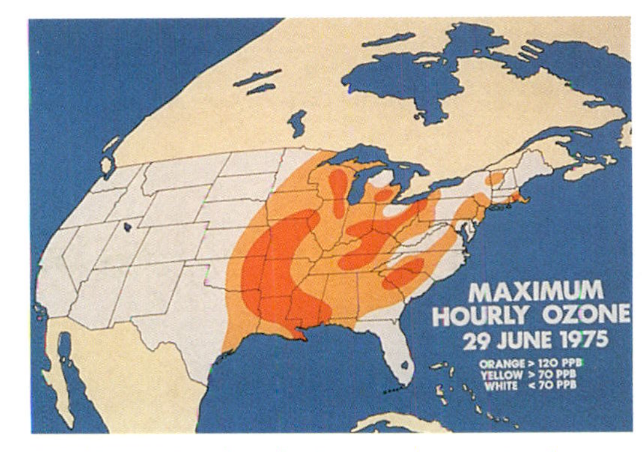

1.2. The darker the shading, the higher the O_3 concentration.

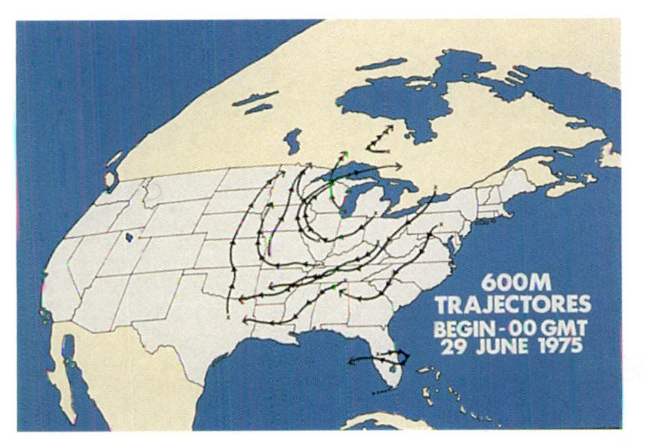

1.3. Two consecutive arrows represent a 12-hr interval in the transport of a given air mass.

1.4. Satellite photograph showing clockwise movement of pollutant clouds from the northeast through the central U.S. into the eastern half of Minnesota, confirming the results shown in Plate 1.3.

2.1. Right, healthy tobacco (*Nicotiana tabacum* cv. Bel-W3) leaf and left, leaf with O_3-induced flecking. (Courtesy W. J. Manning, University of Massachusetts)

2.2. O_3-induced chlorosis and whitish tan interveinal necrosis on the upper surface of a watermelon (*Citrullus lanatus*) leaf. (Courtesy A. S. Heagle, USDA-ARS)

2.3. O_3 injury on a wheat (*Triticum aestivum*) leaf: chlorotic and necrotic white streaks between larger veins. (Courtesy H. E. Heggestad, formerly USDA-ARS)

2.4. O_3-induced bifacial necrosis on pinto bean (*Phaseolus vulgaris*). (Courtesy P. J. Temple, University of California, Riverside)

3.1. Typical acute sulfur dioxide (SO_2) injury symptoms, interveinal whitish brown necrosis, on soybean (*Glycine max*).

3.2. Whitish tan, SO_2-induced acute interveinal necrosis on hollyhock (*Alcea rosea*). (Courtesy D. B. Drummond; reprinted from Krupa et al., 1982)

3.3. Light tannish, SO_2-induced interveinal necrosis on blackberry (*Rubus* sp.).

3.4. Nonchlorophyllous, orange red pigment accumulation (premature senescence) in cotton (*Gossypium hirsutum*) leaves caused by chronic SO_2 exposure. (Courtesy A. S. Heagle, USDA-ARS)

4.1. Lower leaf surface of lettuce (*Lactuca sativa*) with bronzing induced by peroxyacetyl nitrate (PAN).

4.2. The relationship between leaf age and PAN injury on petunia (*Petunia* sp.). Left, youngest leaves show curling of the leaf tips; middle, semimature leaves show the most injury (bleaching resulting in necrosis); and right, most mature leaves show less damage than the semimature leaves.

4.3. Cherry tomato (*Lycopersicon esculentum* var. *leptophyllum*) leaf with interveinal bleaching caused by acute exposure to anhydrous ammonia (NH_3). Note the similarity between these symptoms and those caused by SO_2 (Color Plate 3).

4.4. Gladiolus (*Gladiolus* sp.) leaf with marginal and tip necrosis caused by hydrogen fluoride (HF). (Courtesy A. H. Legge, Biosphere Solutions, Calgary)

5.1. Upper surface of a tulip poplar (*Liriodendron tulipifera*) leaf with O_3-induced purple stipple. (Courtesy J. R. Renfro, U.S. National Park Service)

5.2. Chlorotic mottle on the needles of ponderosa pine (*Pinus ponderosa*). (Courtesy P. R. Miller, USDA-Forest Service)

5.3. O_3-induced, reddish brown tip necrosis and chlorotic mottle on white pine (*Pinus strobus*).

5.4. Scotch pine (*Pinus sylvestris*) with tip necrosis spreading downward, an acute response to high SO_2 concentrations.

6.1. Response of radish to chronic O_3 exposure. Left, healthy plant and right, plant exposed to ambient oxidants (O_3). Note bleaching of the leaves and smaller size of the radish caused by reduced translocation of carbon (photosynthate) from the shoot to the root. This example indicates that effects below ground are as important as those above. (Courtesy H. E. Heggestad, formerly USDA-ARS)

6.2. Comparison of the shoot biomass response of the white clover (*Trifolium repens*) O_3-resistant cultivar NCR (left) and the O_3-sensitive cultivar NCS (right). (Courtesy A. S. Heagle, USDA-ARS)

7.1. Healthy, lush ponderosa pine (*Pinus ponderosa*) in the San Bernardino forest in California in 1963. (Courtesy P. R. Miller; reprinted from Krupa et al., 1982)

7.2. The tree in Plate 7.1 in 1973. Note the defoliation (loss of needles) and loss of tree vigor after long-term stress from photochemical oxidants (O_3) followed by bark beetle attack. (Courtesy P. R. Miller; reprinted from Krupa et al., 1982)

8.1–8.3. Spring onset of new growth in lodgepole pine (*Pinus contorta*) at Whitecourt, Alberta, Canada. The photographs were taken on the same day at three different locations downwind from a SO_2 source. The three sites were ecologically comparable, and the three different trees shown in the photographs were of comparable age but had been chronically exposed to different levels of SO_2. (Courtesy A. H. Legge, Biosphere Solutions, Calgary)

8.1. Control site. The terminal candle of new needles is well open.

8.2. Site at an intermediate distance from the SO_2 source. The terminal candle of new needles is emerging but not open.

8.3. Site close to the SO_2 source. The terminal candle of new needles is not even formed, and older needles are not retained.

the respiratory tract. In the ciliated airways of the nose and upper tracheobronchial zone, clearance in healthy individuals is achieved in less than a day. In general, as the site of deposition becomes more distal, the time required for clearance increases. Clearance of particles in the alveolar region may take weeks to months. Slow clearance of particles from the respiratory system of humans is generally considered to be detrimental, since toxic substances are in contact with sensitive tissue for longer periods of time.

Alveolar deposition of particles 0.1–2.5 μm in diameter is most efficient. The toxicity of these small particles may be greater than that of larger particles, since the concentrations of toxic substances such as lead, zinc, chromium, mercury, sulfates, and nitrates increase with decreasing particle size (see Tables 2.14 and 2.15). Additionally, the enormous surface area of small particles allows for higher reaction and dissolution rates of toxic chemical species. Their relatively long retention in the alveolar region permits substances such as lead to be extracted and transported to other parts of the body.

Particulate matter may contribute to the development of chronic bronchitis and may be a predisposing factor in acute bacterial and viral bronchitis, especially in smokers and children. It may also aggravate bronchial asthma and the late stages of chronic bronchitis and pulmonary emphysema. The health effects of particulate matter may also depend in large measure on synergistic (more than additive) effects with pollutants such as SO_2.

A variety of epidemiological studies have demonstrated an apparent causal relationship between levels of ambient particulate matter and adverse health effects, including 1) excess mortality in the presence of total suspended particulate matter (**TSP**) levels of ~1,000 $\mu g\ m^{-3}$, 2) aggravation of bronchitis at TSP levels of 250–500 $\mu g\ m^{-3}$, and 3) small, reversible changes in the pulmonary function of children at TSP levels of 200–420 $\mu g\ m^{-3}$.

There is a growing body of evidence that acid aerosols, composed primarily of H_2SO_4 and HNO_3, may have detrimental health effects. Animal studies indicate that exposures to acid aerosols at concentrations slightly above ambient levels significantly decrease the ability of the upper respiratory tract and the pulmonary system to remove potentially harmful particles. Acid aerosols at near-ambient concentrations also appear to induce asthmatic attacks in exercising asthmatic adolescents. The incidence of bronchitis among children has also been reported to be correlated with ambient H^+ concentrations.

Lead (Pb)

Since lead is a ubiquitous substance and occurs naturally in small quantities in soil, water, and air, human exposure to a variety of sources and environmental media is inevitable. The lead burden of the human body may result from ingestion of contaminated food and water and inhalation of lead particles. After lead enters the body, it is absorbed into the bloodstream and transported to all parts of the body. Although significant amounts of lead may be found in the blood and soft tissues, it tends to accumulate in the bones, where it is immobilized. It is eliminated from the body primarily by fecal excretion. The actual amount of lead absorbed depends on the form of lead and a person's nutritional status, metabolic activity, and prior exposure history. About 20–40% of inhaled ambient lead particles can be deposited in the lung, where 50% or more is absorbed and enters the bloodstream. Some of the deposited lead particles are removed by pulmonary clearance mechanisms, or they may be swallowed and enter the gastrointestinal tract. In children, about 40% of the lead in particles entering the gastrointestinal tract is absorbed; in comparison, lead absorption in adults through the gastrointestinal system is only about 10%.

In addition to exposure to the lead found in air, water, and food, preschool children may be exposed to significant levels by consuming chips of lead-based interior paints. This is a problem primarily in older, dilapidated housing in inner-city areas. As a consequence of such lead exposure, an estimated 600,000 preschool children in the United States may have blood lead levels sufficient to cause clinical symptoms of acute or chronic poisoning. Because of their compulsion to put objects into their mouths, their exposure to lead sources, and their inherent physiological susceptibility, preschool children are especially vulnerable. The **ambient air quality standard** in the United States for lead (maximum quarterly average of 1.5 $\mu g\ m^{-3}$) is designed to protect primarily this high-risk, critically sensitive population (see Table 2.11).

Emissions of lead to the atmosphere result from the smelting of lead and other heavy metal ores, the combustion of fossil fuels and leaded gasolines, and the manufacture of lead products. Until recently, alkyl lead compounds used as antiknock additives in gasoline were the source of about 90% of all lead entering the atmosphere. With the phasedown of regular leaded gasoline use in the United States, which began in 1974, automobile-related lead emissions to the atmosphere have declined substantially (Fig. 6.3).

Depending on the extent of the exposure, individuals may develop symptoms of acute or chronic lead poisoning. The principal target organs or organ systems include the blood, the brain and nervous system, the kidneys, and the reproductive system. Symptoms from acute exposures may include colic, shock, severe anemia, acute nervousness, kidney damage, and irreversible brain damage. If the dose is sufficiently high, death may result. Chronic poisoning may also result in severe damage to the brain, kidneys, and blood-forming systems.

Hematological changes (i.e., effects on the blood) are the earliest detectable manifestation of low-level, chronic lead exposure. Lead appears to inhibit enzymes in the pathway of hemoglobin biosynthesis. This decreased hemoglobin production is in part the cause of the anemia associated with chronic lead poisoning. At higher doses, anemia may be further aggravated by lead-induced destruction of red blood cells.

In addition to preschool children, women of child-bearing age are at special risk because the potential for deleterious health effects on reproduction and development is high. There is also evidence that lead can cause sterility, abortions, stillbirths, and neonatal (newborn) deaths. Because lead crosses the placenta, high lead levels in the blood of the mother may expose the fetus, resulting in postnatal mental retardation. This problem is of special concern in the lead industries as women enter the work force in these previously all-male occupations.

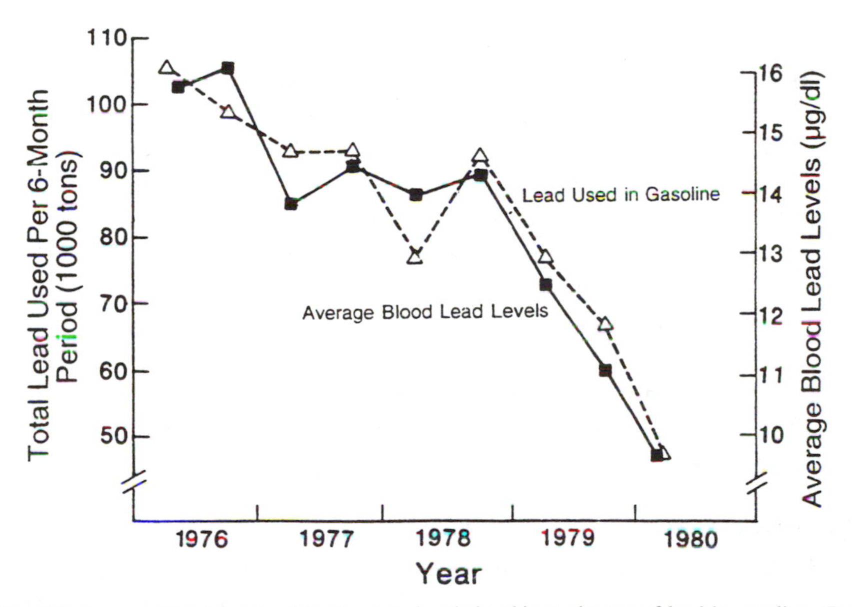

Fig. 6.3. Average blood levels of lead and their relationship to the use of lead in gasoline. (Reprinted, by permission, from Goldsmith, 1986)

Minor Air Pollutants Hazardous to Human Health

The following air pollutants, although hazardous to human health, are not as widespread as the pollutants discussed previously. Therefore, they are classified as minor pollutants. An exception, however, is asbestos.

Arsenic (As)

Arsenic is regulated as a hazardous air pollutant because it can cause lung cancer. The major sources of ambient exposures are primary metal smelters. Because of the high inorganic arsenic contents of some copper ores, arsenic emissions from large primary copper smelters are of particular health concern.

Asbestos

Of the regulated hazardous air pollutants, asbestos is the most prevalent in use and distribution and as a consequence, poses the most significant potential community health problem. Community exposure to asbestos particles occurs from the mining and processing of asbestos ores, manufacturing of asbestos products such as wallboard and insulation, application of asbestos fireproofing materials, demolition of buildings, and abrasion of brake linings and clutch facings. The term "asbestos" is used to describe a number of minerals (e.g., silicon and aluminum), which are fibrous in nature. The inhalation of these fiber fragments has been recognized for some time as the cause of an industrially contracted lung disease called **asbestosis,** which is characterized by a diffuse fibrosis or scarring in the lower lobes of the lungs. It is commonly found in workers 30 years after the initial exposure to asbestos. Asbestos is also known to cause lung cancer and a cancer of the lining of the body cavity called **mesothelioma**. Like asbestosis, asbestos-induced lung cancer and mesothelioma have long latency periods. Once asbestos fibers have entered the body, the potential for cancer exists for the rest of the life of the individual.

The health hazards of asbestos exposure are not limited to asbestos workers. Lung cancer and mesothelioma believed to be caused by asbestos have been recorded in families exposed to the working clothes of those employed in asbestos manufacturing.

Benzene (C_6H_6)

Benzene is a widely used industrial solvent, and more than 2 million workers are exposed to it each day. One of the major collective sources of benzene emissions to the atmosphere is gas stations; benzene is the major additive used in unleaded gasolines to boost octane rating. It is a hematological (blood) poison, causing aplastic (unable to regenerate) anemia (lack of sufficient red blood cells) and acute myelogenous (originating from the spine) leukemia (abnormal increase in leukocytes or white blood cells). The aplastic (incomplete) anemia results from benzene-induced damage to the bone marrow.

Beryllium (Be)

Beryllium is an extremely toxic metal and is used in various industrial products (e.g., railroad tracks). In persons chronically exposed, it produces a granulomatous (granule-textured) inflammation throughout the lung characterized by multiple nodulelike accumulations of cells. In addition, chronic **berylliosis** is characterized by an enlargement of the lymph nodes. As berylliosis begins to develop, it causes a progressive shortness of breath, weight loss, cough, and phlegm production. In later stages, the granulomatous inflammation that is typical of the disease develops, followed by fibrosis and subsequent damage to the heart. One of the principal effects of the disease is interference with the movement of O_2 from the alveoli to the arterial blood. Berylliosis may also affect the kidneys, liver, and spleen, and the skin may be subject to ulcers and granulomas.

In addition to berylliosis, the health hazards of beryllium may also include its ability to cause cancer. Laboratory exposures of animals have shown that it is carcinogenic. However, its ability to cause cancer in humans has not yet been demonstrated.

Although berylliosis is predominantly a disease of those industrially exposed, there is evidence to indicate that individuals residing 0.25–2 miles (0.4–3 km) away from a beryllium processing plant can develop clinical symptoms of chronic beryllium poisoning.

Mercury (Hg)

The nature of mercury poisoning, its symptoms, and the distribution, accumulation, and elimination from the body depend on whether exposure is to elemental mercury or to other mercury compounds. Exposure to mercury in its inorganic forms can damage the liver and kidneys. As mercury vapor diffuses from the lung to the blood, it rapidly reaches the brain, where it interferes with muscular coordination, producing generalized tremors and even hallucinations. Inorganic mercury is eventually excreted from the body through the kidneys. In comparison, the most common effect of exposure to organic mercury compounds such as **methyl mercury** is brain damage characterized by an atrophy or shrinkage of the brain cells. Initial symptoms of organic mercury poisoning include numbness in the fingers, lips, and tongue, slurred speech, **ataxic** (lacking coordination of voluntary muscular movements) gait, difficulties in swallowing, deafness, and blurred vision. In advanced stages, individuals become loud and explosive in their speech and actions. They lose comprehension of what they or others are saying and lose contact with their surroundings (disorientation). Organic mercury compounds such as methyl mercury are also known to cause breaks in chromosomes and result in abnormal cell division.

Vinyl Chloride (VC)

Epidemiological studies suggest that vinyl chloride, a gas used in the manufacture of polyvinyl chloride (PVC) plastics, could cause **angiosarcoma**, a rare form of liver cancer, in workers occupationally exposed. This danger has been confirmed by exposing animals to vinyl chloride.

Vinyl chloride appears to be a **multipotent carcinogen;** that is, it produces cancers in many different organs, and in some cases, it produces different kinds of tumors in the same organ. Angiosarcoma can be induced by exposures to as little as 10 ppm, which is well below the typical exposure levels of workers prior to 1974. In addition to angiosarcoma, vinyl chloride exposures may result in a greater incidence of brain and lung cancer. Several studies have also indicated that vinyl chloride may pose reproductive hazards, because it can cause mutations in bacteria and reduced fertility in laboratory animals exposed to concentrations typical of previously uncontrolled occupational environments.

Pesticides

Although because of regulatory policies, widespread use of pesticides in terms of their absolute quantity in agriculture may have declined in the United States and some other developed countries during the last decade, their use continues unabated in developing countries (UN-FAO, *unpublished*). Nevertheless, with the exception of wind-aided drift, pesticide inhalation or skin contact and absorption are considered to be occupational hazards, particularly among the migrant workers in the United States. Federal laws require pesticide-applicator licensing through testing for their safe use. Pesticides in drinking water and the food chain are a major concern, while atmospheric deposition is considered to have only minor effects on health. However, this is a controversial topic with unproved results.

A Brief Assessment

It has been documented that air pollution episodes of high concentrations can kill people. This drastic effect is, however, dependent upon the toxicological properties of the pollutant in question. Such mass mortality is an acute response to acute exposure. In contrast, chronic effects are long-term responses that are relatively difficult to document. Nevertheless, much of what we know about these effects is based on epidemiological studies; animal studies; and limited, controlled, human-exposure studies. Not only is it extremely difficult to extrapolate laboratory results to the real world, but a single air pollutant does not exist independently. It is a component in a mixture of air pollutants. An air pollutant can produce additive, more than additive, or less than additive effects in combination with others. For example, inhalation of fine-particle acid aerosols is known to substantially increase the effect of O_3 on human health. Scientists disagree on what a "safe level" should be for a given air pollutant. Although it may be impossible to satisfy the entire scientific community, it should be possible to identify a safe level as long as the observed and/or predicted uncertainties are built into it. This means a range of concentrations and exposure durations must be determined and not just a single value.

In some countries (e.g., Canada), air-quality objectives are divided into categories such as "tolerable," "acceptable," and "desirable." This type of division is not practiced in the United States, and therefore identification of an air-quality standard frequently has been a subject of controversy because of the difficulty in establishing broadly accepted exposure (concentration and duration) and response relationships. This subject is discussed further in Chapter 9, Control Strategies for Air Pollution.

References

Finlayson-Pitts, B. J., and Pitts, J. N., Jr. 1986. Atmospheric Chemistry: Fundamentals and Experimental Techniques. John Wiley & Sons, New York.

Godish, T. 1991. Air Quality. 2nd ed. Lewis Publishers, Chelsea, MI.

Goldsmith, J. R. 1986. Effects on human health. Pages 392-463 in: Air Pollution. Vol. 6, Supplement to Air Pollutants, Their Transformation, Transport and Effects. A. C. Stern, ed. Academic Press, New York.

Goodhill, V. 1971. Maimonides—Modern Medical Relevance. XXVI Wherry Memorial Lecture. Trans. Am. Acad. Ophthalmol. Otolaryngol. 75:463-491.

Hinds, W. C. 1982. Aerosol Technology. John Wiley & Sons, New York.

Lippmann, M. 1989. Health effects of ozone: A critical review. J. Air Pollut. Control Assoc. 39:672-695.

National Research Council. 1991. Rethinking the Ozone Problem in Urban and Regional Air Pollution. National Academy Press, Washington, DC.

Stern, A. C., Wohlers, H. C., Boubel, R. W., and Lowry, W. P. 1973. Fundamentals of Air Pollution. Academic Press, New York.

U.S. Department of Health, Education and Welfare. 1969. Air quality criteria for particulate matter. Natl. Air Pollut. Control Adm. Publ. AP-49.

U.S. Environmental Protection Agency. 1988. Review of the National Ambient Air Quality Standards for Ozone—Preliminary Assessment of Scientific and Technical Information. OAQPS Draft Staff Paper. U.S. EPA, Raleigh, NC.

Further Reading

Calabrese, E. J., and Kenyon, E. M. 1991. Air Toxics and Risk Assessment. Lewis Publishers, Chelsea, MI.

Cordasco, E. M., Demeter, S. L., and Zenz, C., eds. 1995. Environmental Respiratory Diseases. Van Nostrand Reinhold, New York.

Lipfert, F. W. 1994. Air Pollution and Community Health: A Critical Review and Data Sourcebook. Van Nostrand Reinhold, New York.

National Institutes of Health. 1993. Environmental Epidemiology. National Institute of Environmental Health Sciences, Research Triangle Park, NC.

Witorsch, P., and Spagnolo, S. V., eds. 1994. Air Pollution and Lung Disease in Adults. CRC Press, Boca Raton, FL.

CHAPTER 7

Air Quality and Crops

> This coale . . . flies abroad . . . and in the springtime besoots all the leaves, so as there is nothing free from its universal contamination . . . and kills our bees and flowers abroad, suffering nothing in our gardens to bud, display themselves, or ripen.
>
> —John Evelyn; quoted by M. Treshow

Introduction

Although this chapter deals specifically with crops, in order to provide continuity and appropriate comparison, references to the responses of tree species are made in some tables and notes. However, Chapter 8, Air Quality and Forests, deals with the other aspects regarding trees.

Crops and other higher plant species are stationary, as opposed to some lower plant species, animals, and people, which are mobile. In spite of this, there is a great deal of heterogeneity among crops at the genus, species, and variety or cultivar levels in their responses to air pollution. Frequently, this is true even with clonally propagated species. As with humans, a crop's response to air pollution stress can be **acute** (i.e., response to short-term pollutant episodes), frequently manifested by foliar injury symptoms (Color Plates 2–5 and Notes 7.1), or **chronic** (i.e., response to long-term exposures of relatively low pollutant concentrations and periodic, intermittent, relatively high concentrations). Chronic responses may or may not be accompanied by foliar injury symptoms, yet can result in reduced growth and productivity (yield) losses.

Gaseous air pollutants enter leaves primarily through the stomata (openings on the lower surface of the foliage of most plants through which gas exchange and transpiration occur), although some pollutants such as SO_2 can enter to a lesser degree through the cuticle. Leaves themselves also release gases (e.g., CO_2 through respiration and hydrogen sulfide, H_2S, from the reduction of internal tissue sulfate, SO_4^{2-}).

It is important to note that some atmospheric chemical constituents such as sulfur and nitrogen species are essential crop nutrients or elements. However, depending on their molecular form and concentration (including the duration of occurrence in the atmosphere), they can also be toxic (e.g., SO_2). Although we are concerned here with the atmospheric input of sulfur and nitrogen species, it should be noted that soil serves as a dominant source of these major essential elements. Problems start when the soil supplies the crop's requirement and the atmosphere overloads it (Fig. 7.1).

Notes 7.1. Definitions of some technical terms used in air pollution–plant effects literature

Adventitious root	=	Roots growing from an abnormal position, e.g., from the bottom portion of the stem, above the surface of the soil.
Banding	=	Bands of yellow tissue alternating with green tissue, on conifer needles.
Bifacial	=	Both sides of the leaf.
Bronzing	=	Bronze coloring on the upper (O_3) or lower (PAN) surface of the leaf.
Chlorosis	=	Loss of chlorophyll or yellowing.
Epinasty	=	A plant part (e.g., petiole) is moved outward and downward like a shepherd's crook.
Fleck	=	A spot or a mark usually white, grey or brown with the tissue becoming dry or elevated above the surface of the leaf (on the upper leaf surface).
Interveinal	=	Between veins.
Monocot	=	Plants with a single cotyledon, as opposed to others with two.
Mottle	=	Blotches of yellow color (chlorosis) mixed with the normal green surface.
Necrosis	=	Death of tissue, usually brown, dark brown, red or black in color.
Silvering	=	Turning color to silver.
Stipple	=	Dotted non–green areas on the leaf (on the upper leaf surface).

Thus, air pollution can affect a plant directly by damaging the foliage or canopy (e.g., O_3), indirectly through its effects on the soil (e.g., trace metals), or in both ways (e.g., sulfur). Although most studies have dealt with the effects of pollution on crop growth and productivity, some air pollutants, such as trace metals and organic compounds, can accumulate

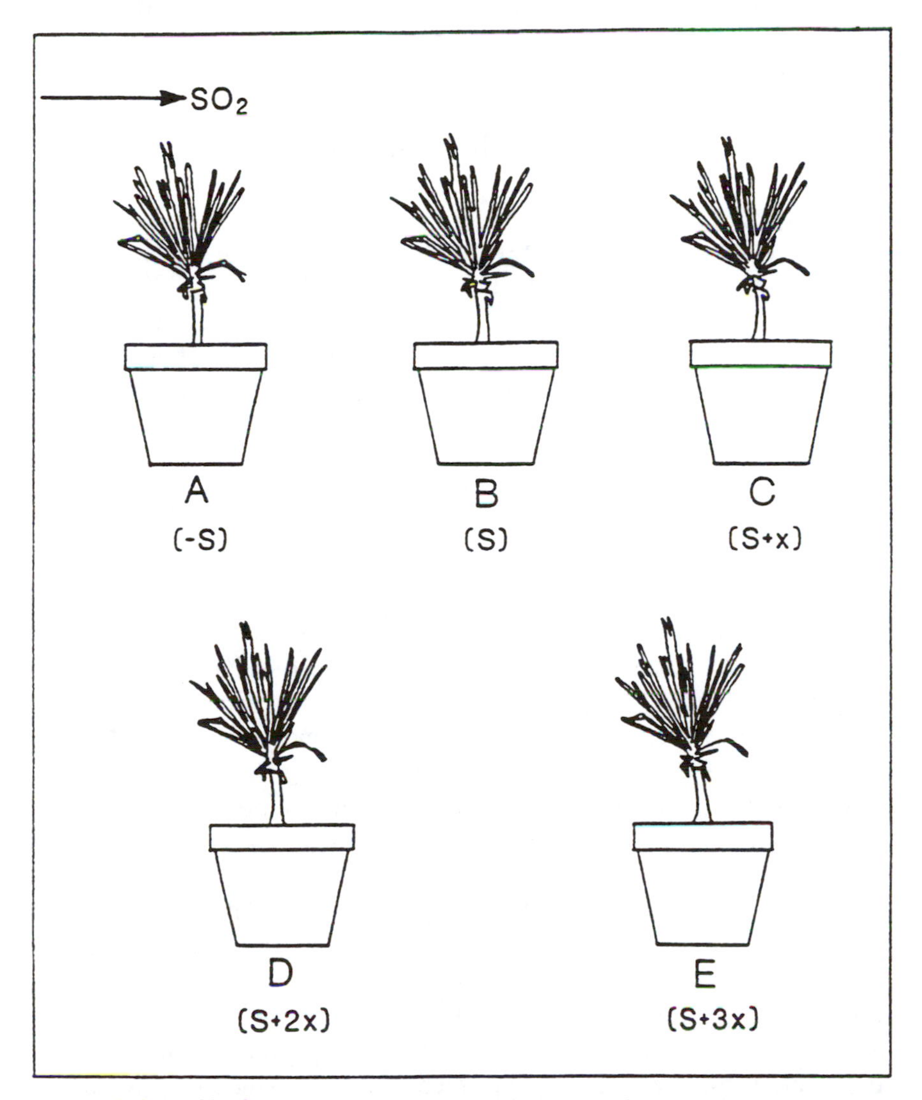

Fig. 7.1. Relationships between soil sulfur content (as a source of plant sulfur requirement) and plant response to atmospheric SO_2. **A,** Sulfur-deficient soil: atmospheric SO_2 can stimulate plant growth by providing the needed sulfur for the plant. **B,** Sulfur-sufficient soil: atmospheric SO_2 may or may not cause negative plant effects depending on the exposure. **C–E,** Increasing amounts of excess soil sulfur for the plant: increasing degree of atmospheric SO_2 injury because of increasing level of plant sulfur saturation from the soil.

in edible crop parts and play a significant role in the food chain. For example, accumulation of fluoride in tea leaves (a known accumulator of fluoride) can lead to the loss of enamel in the teeth of individuals who drink too much tea over a long period (years). This has been observed among poor people or laborers in the Middle to Far Eastern countries. This subject is examined further in the section on hydrogen fluoride in this chapter.

Effects of Select Air Pollutants on Crops

Ozone (O_3)

Ozone is all pervasive at the surface and is the most important **phytotoxic** (toxic to plants) air pollutant (Krupa and Manning, 1988). In contrast to sulfur pollution, which dates back to the Roman Empire (see Chapter 3), the importance of O_3 (a key component of photochemical smog) as a phytotoxic air pollutant was first recognized during the 1950s through studies conducted in the Los Angeles area.

Since those early investigations, it has become evident that O_3 is the most important phytotoxic air pollutant in the United States and elsewhere, causing foliar injury to many agronomic and horticultural crops, deciduous trees, and conifers (Krupa and Manning, 1988). Ozone is an all-pervasive air pollutant. Thus, particularly since the 1960s as urban centers and industries have continued to grow, an increasing number of reports have appeared in the literature regarding O_3-induced foliar injury on sensitive plants in countries such as Australia, Canada, Germany, Great Britain, Greece, India, Israel, Italy, Japan, Mexico, the Netherlands, Poland, Spain, and the Ukraine (Fig. 7.2).

Since surface-level O_3 is primarily produced by **photochemical** processes, high concentrations of O_3 in the atmosphere are related to high radiation and temperature and low relative humidity (O_3 decomposes rapidly at high humidity and low acidity and essentially drops to zero concentration during rainfalls). Thus, high atmospheric O_3 concentrations are normally observed during the warm season or summer months (also the crop growth season). Similarly, during that period in agricultural areas, high O_3 concentrations frequently occur during the early to late afternoon or early evening (see Fig. 2.3). Much of this pattern coincides with the daily periods of high CO_2 (carbon dioxide) uptake or assimilation or photosynthesis by crops and other plants. Thus, unless soil moisture is limiting, the stomata remain open. As previously stated, O_3 enters the leaf through stomata, and its primary site of injury is the

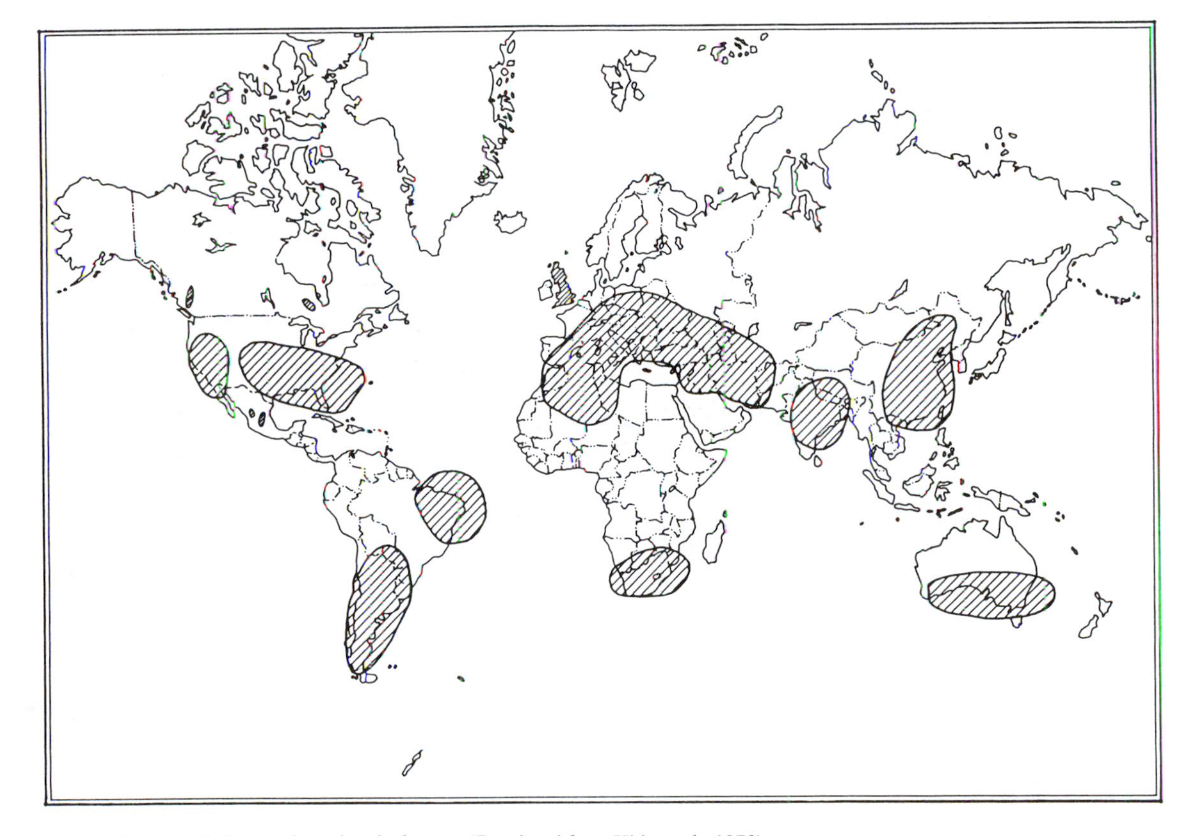

Fig. 7.2. Regions of high susceptibility to photochemical smog. (Reprinted from Hidy et al., 1978)

cells of the **mesophyll** or **palisade layer** on the upper leaf surface. This tissue contains chloroplasts (chlorophyll-containing plastids) and is responsible for photosynthesis or carbon dioxide assimilation. Therefore, a primary effect of O_3 is on photosynthesis, and exposures to sufficient O_3 can produce a variety of injury symptoms associated with chlorophyll breakdown (Tables 7.1–7.3). It is very important to note that these injury symptoms may or may not be associated with growth and yield reductions, unless they occur repeatedly during the growth season. Conversely, growth and yield reductions can occur without or in the absence of foliar injury symptoms. Therefore, crop yield losses caused by O_3 under field conditions are frequently not detected. However, there are experimental methodologies available to document such effects (Manning and Krupa, 1992). Through these studies, it has been shown that particular cultivars or clones of certain sensitive crops such as wheat (Table 7.4), soybean, radish, and clover may exhibit anywhere from 10 to 40% yield loss during a particular year and at a particular location in response to experimental O_3 exposures in the United States (Color Plate 6). These types of yield reductions result from chronic O_3 exposures, and in examining these losses, considerations such as the dynamics of O_3 concentrations and their timing are extremely critical.

Table 7.1. Crops commonly affected by ozone and typical symptoms expressed[a]

Plant (genus)	Foliar symptoms
Bean (*Phaseolus* sp.)	Bronzing; chlorosis; bifacial isolated necrosis of small areas
Cucumber (*Cucumis* sp.)	White stipple
Grape (*Vitis* sp.)	Red to black stipple
Morning glory (*Ipomoea* sp.)	Chlorosis
Onion (*Allium* sp.)	White flecks; tip dieback
Potato (*Solanum* sp.)	Gray fleck; chlorosis; bronzing
Soybean (*Glycine* sp.)	Red bronzing; chlorosis; purple stipple
Spinach (*Spinacia* sp.)	Gray to white fleck
Tobacco (*Nicotiana* sp.)	Metallic to white fleck
Watermelon (*Citrullus* sp.)	Gray fleck

[a] Modified from Krupa and Manning, 1988.

Table 7.2. Deciduous trees commonly affected by ozone and typical symptoms expressed[a]

Tree species	Foliar symptoms
Black cherry (*Prunus serotina*)	Red black stipple; reddening and leaf chlorosis; premature defoliation
Green ash (*Fraxinus pennsylvanica* var. *lanceolata*)	Red purple stipple
Quaking aspen (*Populus tremuloides*)	Black stipple; chlorosis; premature defoliation
Sycamore (*Platanus occidentalis*)	Chlorosis; early senescence (fall coloring) and defoliation
Tulip poplar (*Liriodendron tulipifera*)	Dark stipple (classic symptoms)

[a] Modified from Krupa and Manning, 1988.

For example, a soybean crop may be significantly damaged by hail during the early part of the growth season, and yet there may be no crop loss at harvest because the crop has had sufficient time to recuperate, repair the damage, and compensate for the stress. On the other hand, if the same hail damage occurred during the flowering or pod-filling stages, it could lead to substantial yield reduction. Because of the year-to-year variability in the climate and in O_3 production, crop yield responses also vary from year to year. Ozone-induced crop losses in the United States have been estimated to be approximately \$2–4 billion per year. However, this estimate is based on the assumption that crop stress is caused by O_3 alone under ideal growth conditions. Realistically, much more crop loss can occur because of the prevalence of many other stress factors such as pests, flooding, and drought. Nevertheless, the preceding narrative provides a rationale for concern about surface-level O_3 and crop production.

A number of sensitive cultivated and native plant species respond to O_3 episodes rapidly within days by displaying typical O_3 symptoms on their foliage (Tables 7.1 and 7.2). Plant scientists have taken advantage

Table 7.3. Conifers commonly affected by ozone and typical symptoms expressed[a]

Tree species	Foliar symptoms
Eastern white pine (*Pinus strobus*)	Chlorotic fleck or mottle on older needles; red brown tipburn of current needles
Jeffrey pine (*P. jeffreyi*)	Chlorotic mottle of older needles (banding possible)
Ponderosa pine (*P. ponderosa*)	Chlorotic fleck or mottle on older needles followed by needle dieback from tips (banding possible)
White fir (*Abies concolor*)	Chlorotic mottle on older needles

[a] Modified from Krupa and Manning, 1988.

Table 7.4. Effects of ozone on yields of the winter wheat cultivar Vona in open-top chambers in Ithaca, New York[a]

	Seed weight		100-seed weight	
Treatment[b]	**Weight ($kg\ ha^{-1}$)**	**Loss (%)**	**Weight (g)**	**Loss (%)**
Filtered air	5,331.0	...	3.26	...
Ambient air	4,049.8	24	2.32	29
Nonfiltered air	3,552.1	33	2.47	24
+ 0.03 ppm O_3	2,322.3	56	1.77	46
+ 0.06 ppm O_3	1,698.8	68	1.41	57
+ 0.09 ppm O_3	1,430.0	73	1.30	69

[a] Modified from Kohut et al., 1987.

[b] **Filtered air** = ambient air filtered through particulate filter + activated charcoal. This process usually removes the dust and more than 70% of the ozone in the air. This serves as a control in experiments. **Nonfiltered air** = ambient air filtered for dust only. This procedure removes about 10% or more of ozone in the air. **Nonfiltered + (*x*) ppm O_3** = nonfiltered air to which a known amount of ozone is artificially added, resulting in ozone concentrations above the ambient levels.

of this by using sensitive plant species, such as appropriate tobacco and bean cultivars, as **biological indicators** of relative O_3 pollution in comparisons of the air quality at different geographic locations during a given year and during different years at a given location (Manning and Feder, 1980; Krupa et al., in press). This general concept of using bioindicators to assess relative air quality is a unique feature for studies involving higher plants and even some lower plants such as the lichens but has no comparable application in human health studies, with the exception of mutagenicity tests (refer to the section on hydrocarbons in Chapter 6). It should be pointed out that bioindicator systems are available for water pollution studies, and similar approaches are being developed for the assessment of overall ecosystem health.

Sulfur Dioxide (SO_2)

While O_3 is a secondary pollutant and its adverse effects on crops are problematic at the regional, national, and international scales, SO_2 is of concern primarily in the vicinity of its sources. High SO_2 concentrations occur at ground level during surface-level inversion conditions (see Fig. 1.2), increasing the potential for acute effects on plants.

As previously stated, SO_2 enters the leaf primarily through the stomata, although up to approximately 15% can enter the leaf directly through the cuticle. This is a very important point. While deciduous plants shed their leaves during the fall, evergreens retain their foliage and thus can absorb SO_2 during the winter. Sulfur dioxide is an accumulative poison; in other words, plants exposed to chronic SO_2 pollution accumulate sulfur in the foliage. Thus, evergreens that have accumulated sulfur during the winter become even more vulnerable to spring and summertime SO_2 pollution. This is not to say that deciduous plants are not vulnerable. The sensitivity of a plant is dependent upon the rate of absorption of SO_2 during its active growth, and this varies at the genus, species, and even cultivar levels. As previously stated, sulfur is an essential element, and therefore how a plant will respond to SO_2 exposures depends on whether or not its sulfur requirement is satisfied (Fig. 7.1). Table 7.5 provides a summary of typical SO_2 injury symptoms on both broad-leaved plants and conifers, and Table 7.6 lists species known to be sensitive to acute SO_2 exposures.

Once SO_2 enters the leaf, it is converted first to the **toxic bisulfite** (HSO_3^-) and **sulfite** (SO_3^-) ions and later to the **nontoxic sulfate**

(SO_4^{2-}) ions. If the rate of HSO_3^- and SO_3^- accumulation is faster than that of its conversion to SO_4^{2-}, then injury will occur. On the other hand, if conversion to SO_4^{2-} is faster than accumulation of HSO_3^- and SO_3^-, then injury will not occur.

Unlike O_3, SO_2 has adverse effects on crops in the vicinity of its sources (Table 7.7). Before the 1980s, this was a concern in the United States and other developed nations. In more recent years, because of air quality regulations and application of control technologies (see Chapter 9), acute SO_2 injury to crops is uncommon. However, this does not mean that chronic response (growth and yield effects without symptoms) is not a concern. Refer to the discussion in Chapter 4 on the role of SO_2 in the occurrence of acidic rain. Also of concern are the joint effects of SO_2 and other air pollutants, which are discussed in a later section of this chapter.

Table 7.5. Acute foliar injury symptoms caused by some air pollutants on broad-leaved plants and conifers

	Foliar symptoms	
Air pollutant	**Broad-leaved plants**	**Conifers**
Ammonia (NH_3)[a]	Interveinal chlorosis; necrosis; bleaching; defoliation	Tip necrosis spreading downward
Chlorine (Cl)[a] and hydrogen chloride (HCl)[a]	Bleaching; necrosis; defoliation	Tip necrosis spreading downward
Ethylene (C_2H_4)[a]	Chlorosis; epinasty; premature senescence; adventitious root formation; dry sepal condition of the flower; premature opening of buds; inhibition of bud opening and abscission	Not well studied
Hydrogen fluoride (HF)	Marginal and tip chlorosis or necrosis spreading inward (bifacial)	Tip necrosis spreading downward
Nitrogen dioxide (NO_2)	Waxy coating of the lower surface; brown to rust-colored bifacial interveinal necrosis (not common under most ambient conditions)	Not understood
Particulate matter (PM)	Chlorosis; encrustation of PM; abscission	Tip chlorosis and necrosis spreading downward
Peroxyacetyl nitrate (PAN)	Bronzing and silvering of the lower surface; also bifacial bleaching; banding on monocots	Relatively resistant; not well studied
Sulfur dioxide (SO_2)	Interveinal chlorosis or necrosis (bifacial)	Tip necrosis spreading downward

[a] Frequently occurs at high concentrations for a few to several hours because of accidents or equipment malfunction. Therefore, injury symptoms occur on many plant species over a short duration. Plant response is easily related to the occurrence of the accident or equipment malfunction through proper retrospective investigation.

Table 7.6. Some plant species known to be sensitive to acute injury by ethylene, hydrogen fluoride, peroxyacetyl nitrate, or sulfur dioxide

Ethylene (C_2H_4)	**Hydrogen fluoride (HF)**	**Peroxyacetyl nitrate (PAN)**	**Sulfur dioxide (SO_2)**
Carnation (*Dianthus caryophyllus*)	Blueberry (*Vaccinium* sp.)	Bean (*Phaseolus* sp.)	Alfalfa (*Medicago sativa*)
Cucumber (*Cucumis sativus*)	Box elder (*Acer negundo*)	Romaine lettuce (*Lactuca* sp.)	Barley (*Hordeum vulgare*)
Easter lily (*Lilium longiflorum*)	Chinese apricot (*Prunus* sp.)	Swiss chard (*Beta vulgaris*)	Bean, field (*Phaseolus* sp.)
Geranium (*Geranium* sp.)	Corn, sweet (*Zea mays*)	Tomato (*Lycopersicon esculentum*)	Eastern white pine (*Pinus strobus*)
Honey locust (*Gleditsia triacanthos*)	Douglas fir (*Pseudotsuga menziesii*)	White-flowered petunia (*Petunia* × *hybrida*)	Green ash (*Fraxinus pennsylvanica*)
Marigold (*Tagetes* sp.)	Eastern white pine (*Pinus strobus*)		Jack pine (*Pinus banksiana*)
Orchis (*Orchis* sp.)	European grape (*Vitis* sp.)		Maple (*Acer* sp.)
Pea (*Pisum sativum*)	Gladiolus (*Gladiolus* sp.)		Oats (*Avena sativa*)
Rose (*Rosa* sp.)	Italian prune (*Prunus* sp.)		Paper birch (*Betula papyrifera*)
Tomato (*Lycopersicon esculentum*)	Mugho pine (*Pinus mugo*)		Red pine (*Pinus resinosa*)
	Ponderosa pine (*Pinus ponderosa*)		Rye (*Secale cereale*)
	Scotch pine (*Pinus sylvestris*)		Soybean (*Glycine max*)
	Tulip (*Tulipa* sp.)		Trembling aspen (*Populus tremuloides*)
	Western larch (*Larix occidentalis*)		Wheat (*Triticum aestivum*)
			White birch (*Betula* sp.)

Table 7.7. Some crop species that have exhibited acute foliar injury under ambient conditions caused by sulfur (SO_2) contaminants from industrial point or source complexes

Location	Source	Species with acute injury	Reference
North America			
Sudbury area, Ontario, Canada	Copper and nickel smelter complex	Buckwheat (*Fagopyrum esculentum*), barley (*Hordeum vulgare*), red clover (*Trifolium pratense*), radish (*Raphanus sativus*), oats (*Avena sativa*), pea (*Pisum sativum*), rhubarb (*Rheum rhabarbarum*), timothy (*Phleum pratense*), swiss chard (*Beta vulgaris* subsp. *cicla*), bean (*Phaseolus vulgaris*), beet (*Beta vulgaris*), turnip (*Brassica rapa*), carrot (*Daucus carota*), cucumber (*Cucumis sativus*), lettuce (*Lactuca sativa*), tomato (*Lycopersicon esculentum*), potato (*Solanum tuberosum*), raspberry (*Rubus* sp.), celery (*Apium graveolens*), spinach (*Spinacia oleracea*), cabbage (*Brassica oleracea*), corn (*Zea mays*)	Dreisinger and McGovern (1970)
Widow's Creek, Alabama, United States	Coal-fired electricity generating plant	Soybean (*Glycine max*)	McLaughlin and Lee (1974)
Douglas/Hereford, Arizona, United States	Copper smelter	Lettuce (*Lactuca sativa*), green onion (*Allium cepa*), cabbage (*Brassica oleracea*), carrot (*Daucus carota*), chili pepper (*Capsicum annuum*), cotton (*Gossypium hirsutum*)	Haase et al. (1980)
An area in southeastern Ohio and northwestern Virginia, United States	Coal-fired electricity generating plant	Bush bean (*Phaseolus* sp.), corn (*Zea mays*)	Jacobson and Showman (1984)
Europe			
Biersdorf, former West Germany	Metal smelter	Winter wheat (*Triticum aestivum*), spring wheat (*Triticum aestivum*), oats (*Avena sativa*), winter rye (*Secale cereale*), potato (*Solanum tuberosum*)	Guderian and Stratmann (1962)

Hydrogen Fluoride (HF)

Hydrogen fluoride is a primary, gaseous pollutant. Like SO_2, it is an accumulative poison. Most soils contain some soluble fluoride, and thus most plants contain a background concentration of <10 ppm in their foliage. Some plants, such as tea and dogwood, are natural accumulators of fluoride (up to 200 ppm in their foliage). Therefore, the consequence of drinking too much tea and its effects on teeth were mentioned previously.

A summary of acute HF injury symptoms on plants and a list of sensitive plant species are given in Tables 7.5 and 7.6, respectively. In addition to foliar injury, fluoride can damage fruits. For example, on peaches, it causes a problem known as **suture** (a seamlike joint on the fruit) **soft spot** that makes the fruit unmarketable.

Gladiolus is extremely sensitive to HF, and accumulation of fluoride in its foliage (as little as 20 ppm) can result in injury symptoms; in most other crops, accumulation levels of 30–100 ppm are required. Hydrogen fluoride is probably the only air pollutant for which a direct correlation exists between its atmospheric concentration (a few parts per billion continuously throughout the entire growth season) and its foliar concentration (many parts per million). In comparison, with SO_2 such a relationship is not simple (i.e., it is highly nonlinear).

Fluoride is a respiratory poison and is very reactive. Since it reacts even with glass, hydrofluoric acid, which is used to etch glass, is stored in plastic containers. In addition to rotting human teeth when excessive amounts of tea are consumed over many years, fluoride causes **fluorosis** in animals (e.g., cattle) that are fed hay or fodder high in fluoride. As in humans, initial symptoms of fluorosis in animals involve **mottling** or loss of tooth enamel (at ~500 ppm oral fluoride) followed by softening of the bones (at >2,000 ppm bone tissue fluoride), weakening of the animal (e.g., a cow), and finally death. Obviously, this is paralleled by a reduction in milk production. While high fluoride levels can be found in milk, levels in meat are undetectable.

Since it is relatively easy to control, atmospheric fluoride at the present time is not a common problem in the United States, although it occurs in both particle and gas phases. Fluoride reacts very rapidly with alkali (see Chapter 9), and it is used in an emission-control strategy. Water is fluoridated at low concentrations as an aid to human dental hygiene, but its use in irrigating residential or greenhouse plants can result in injury to ornamental or horticultural species (e.g., in Florida), particularly if the plants are grown in fluoride-rich soils.

Peroxyacetyl Nitrate (PAN)

In liquid form, PAN is explosive when agitated. Fortunately for us, it exists only as a gas in the atmosphere. It must first be produced in a liquid phase to obtain sufficient quantities of the gas, and because of its explosiveness, few laboratories produce PAN for plant-exposure studies. Therefore, experimental data on PAN exposures and crop responses are limited. Nevertheless, ambient PAN concentrations frequently occur at about 10–20% (<20 ppb) of the corresponding O_3 concentrations. Yet they can cause injury symptoms on sensitive plant species (Tables 7.5 and 7.6).

Peroxyacetyl nitrate is a very interesting air pollutant. To cause injury symptoms on plants, light is required prior to, during, and after exposure. If light is absent during any of these periods, symptom production will be suppressed. Unlike O_3, which produces typical symptoms on the upper leaf surfaces, PAN produces injury symptoms on the lower leaf surfaces (exceptions exist, e.g., the white-flowered varieties of petunia, on which symptoms, bleaching, and necrosis are produced on both sides). PAN accumulates in the spaces between the spongy parenchyma cells of the lower leaf surface, although the exact mechanism of injury is not known. In contrast to some ornamental (e.g., petunia) and agronomic (e.g., bean) species, trees appear to be relatively resistant to PAN. However, these observations are based on a very limited number of studies. More importantly, PAN is produced through photochemical processes similar to those that produce O_3, and the joint effects of these two pollutants are not well understood.

Ethylene (C_2H_4)

Ethylene is a minor pollutant of local concern (see Table 2.1). Yet it can be of serious economic consequence for the greenhouse industry (from improper venting of the exhaust from natural gas or oil heaters). During the 1960s and 1970s, ornamental industries in the United States lost several million dollars' worth of orchids in California and Easter lilies on the East Coast because of ethylene incidents.

Ethylene is a natural plant growth hormone, and yet concentrations (in parts per billion) of C_2H_4 in the atmosphere for several hours can result in injury to foliage and flowers on sensitive ornamental and horticultural crops (Tables 7.5 and 7.6). Like O_3 and PAN, C_2H_4 is not an accumulative poison. Yet it is very effective at parts-per-billion levels in the air, particularly in closed environments such as greenhouses. Tomato and pea are excellent biological indicators for C_2H_4 pollution.

Lead (Pb), Trace Metals, and Organic Pollutants

Particulate lead, trace metals (e.g., cadmium), and organic pollutants (e.g., polyaromatic hydrocarbons or PAHs) can accumulate in plant parts without inducing any injury symptoms. Such accumulation can occur both through the air and through the soil. Humans and animals that consume such plant material may be adversely affected. Classical lead poisoning of humans from the consumption of wine served from lead vessels has been documented as far back as the Roman Empire. More recently, the use of lead in gasoline and its accumulation in urban vegetable gardens near highways has been of concern. Classic lead poisoning of cattle near a battery-recycling plant in Minnesota was one of the first reports on that subject. Trace metal accumulation in crops can affect their responses to other pollutants (e.g., SO_2). However, this is not a well-researched subject. The same is true for the accumulation of organic pollutants. Overall, the role of contaminant accumulation (other than fluoride) in edible plant parts, its consumption, and the impacts on the food chain need extensive examination. At the present time, the subtlety and complexity of this subject has not attracted as much attention as the other, more directly researchable aspects of air pollution.

Crop Responses to Multiple Air Pollutants

Ambient air is always composed of pollutant mixtures, although the level of a particular air pollutant may be higher than the levels of others at a particular time and location. Under these conditions, typical foliar injury symptoms can be seen on crops in response to the pollutant that occurs in large enough toxic concentrations. In contrast, as previously noted, chronic exposures can result in crop yield reductions without symptoms. The joint effects of pollutant mixtures in this regard are poorly understood. The main problem is the lack of sufficient field research, which is extremely expensive to do. There are indications, however, that the joint effects of two or more pollutants can be additive ($2 + 2 = 4$), more than additive ($2 + 2 = 6$), or less than additive ($2 + 2 = 3$). Such effects vary at the genus, species, and cultivar levels. Furthermore, such effects are dependent upon

1. the physical and chemical characteristics of the individual pollutants in the mixture;
2. the sequence in which crops are exposed to high concentrations of the individual pollutants (e.g., high NO_2 concentrations occur dur-

ing the early and late parts of the day while high O_3 concentrations occur during the middle to late part of the day);

3. the relationship between time of the day when a given pollutant occurs at high concentrations and time of the day when crops are most sensitive;
4. the ratio or ratios of pollutant concentrations in the mixture;
5. the growth stage of the crop; and
6. other environmental factors.

These complexities further contribute to our inability to make generalizations. Nevertheless, limited available data provide support to the type of joint effects previously described.

Rain must also be considered to represent mixtures of pollutants (see Chapter 4). Acidic rain has been considered a serious environmental threat. Interestingly, there is no single published evidence to document adverse effects of ambient acidic rain on crops under field conditions. A large amount of literature exists on the negative effects of **SAR** or **simulated acidic rain** on crops under experimental conditions. However, these studies are highly controversial because they 1) represent unreal exposures to very high acidity; 2) subject crops to exposures that are not similar to the occurrence or patterns of ambient rain at a given location; and 3) optimize the experimental conditions to provoke the worst-case scenario. These types of studies have caused credibility problems within the scientific community both in the United States and Europe, and at the present time, this type of research is declining significantly.

Crop Responses to Global Climate Change

The history of the development of the world's agricultural systems is to some extent the history of climatic changes of the past coupled with technological advances that have permitted the successful production of crops and livestock in regions of the world far removed from their biological origins. However, such developed agricultural systems lack much of the homeostatic resilience of natural ecosystems and therefore have limited ability to resist climate change. Table 7.8 provides a summary of known plant responses to elevated levels of CO_2, UV-B radiation, and O_3. For detailed information, the reader is referred to Krupa and Kickert (1989), Runeckles and Krupa (1994), and Rogers et al. (1994). Because of the types of crop responses outlined, there will

Table 7.8. Overview of the general effects of CO_2, UV-B, and O_3 on crops in single-exposure mode[a,b]

	Crop response to environmental change		
Plant characteristic	**Doubling of CO_2 only[c] (direct effect)**	**Increased UV-B only (stratospheric O_3 depletion)**	**Increased tropospheric O_3 only**
Crop maturation rate	Increases	Not affected	Decreases
Drought stress sensitivity	Plants become less sensitive to drought	Plants become less sensitive to UV-B but sensitive to lack of water	Plants become less sensitive to O_3 but sensitive to lack of water
Dry matter production and yield	Nearly doubles in C_3 plants, but C_4 plants show only small increases	Decreases in many plants	Decreases in many plants
Flowering	Earlier flowering	Inhibits or stimulates flowering in some plants	Decreased floral yield and number and yield of fruits, and delayed fruit setting
Mineral stress sensitivity	Plants become less responsive to elevated CO_2	Some plants become less sensitive while others become more sensitive to UV-B	Plants become more susceptible to O_3 injury
Photosynthesis	Increases in C_3 plants up to 100%, but C_4 plants show only a small or no increase	Decreases in many C_3 and C_4 plants	Decreases in many plants
Sensitivity between species	Major differences between C_3 and C_4 plants	Large variability in response among species	Large variability in sensitivity among species
Sensitivity within species (cultivars)	Can vary among cultivars	Response differs among cultivars of a species	Response differs among cultivars of a species
Water-use efficiency	Increases in C_3 and C_4 plants	Decreases in most plants	Decreases in sensitive plants

[a] Modified from Krupa and Kickert, 1989.

[b] There are exceptions to these general statements. CO_2 = carbon dioxide; O_3 = ozone; and UV-B = radiation at 280 to 320 nm.

[c] C_3 = almost all temperate crops, in which the assimilated CO_2 is fixed in a molecule made of three carbon atoms. C_4 = crops of tropical origin, such as sorghum and sugarcane, in which the assimilated CO_2 is fixed in a molecule made of four carbon atoms.

inevitably be changes in agricultural systems as average values and standard deviations of temperatures and rainfall patterns shift and as changes occur in tropospheric levels of atmospheric trace gases and irradiation.

Dekker and Achununi (1990) point out that all sectors of agriculture are sensitive to weather and climate and any substantial change will alter total productivity. However, this does not necessarily imply reduced productivity. In developed areas, agriculture is supported by a complex system of research, education, information, finance, and farm supply overlaying the agricultural potentials of the available soils. Global warming will eventually result in a decoupling of soil from climate, because although soil characteristics are partly the result of past climates, soil properties change slowly relative to climatic change. Public policy and agricultural management will attempt to develop strategies for maintaining production in areas with the best soils, in spite of shifts in climate. For example, Dekker and Achununi (1990) cite the shift from wheat to barley in the northern Canadian wheat belt attributable to shortened growing seasons. In contrast, the growth in U.S. soybean acreages of the past has been almost entirely driven by the prospects of increased profitability.

Of equal or greater importance than the direct effect of rising temperatures is the indirect effect on the hydrologic cycle, leading to shifts in the dependence on irrigation, where such water is available. Projected drier summers in parts of the U.S. corn belt will probably result in a shift from the production of corn to grain sorghum (Dekker and Achununi, 1990).

Provided that climate change is gradual, plant breeders knowingly or unknowingly (while screening cultivars for other reasons) are likely to develop new cultivars better suited to the altered conditions. However, as cropping practices change, parallel changes in pest control and management must be made in anticipation of the upsurge of pests and diseases well suited to the changed conditions. At least in some countries, biological control of pests and diseases (including the use of resistant crop varieties and cultivars) is now the main alternative used to minimize environmental contamination by chemical agents. Such an approach, however, has proved to be slow but is an extremely important consideration in assessing the impacts of climate change on agriculture. In sustainable agriculture, minimum energy input and consistent energy output should be major considerations. Inasmuch as evolving climate and human intervention have shaped our agriculture, there are lessons to be learned from historical crop-management practices.

Although CO_2 is a prime factor in projected global warming, increased CO_2 levels in the troposphere will influence crop production directly because of its effects on photosynthesis (Rogers et al., 1994). Although **C_4 plants** (plants that assimilate CO_2 into an organic molecule containing four carbon atoms) such as corn, sorghum, millet, and sugarcane are more efficient users of CO_2 than **C_3 species** (plants that assimilate CO_2 into an organic molecule containing three carbon atoms) such as wheat, rice, and almost all other crops, C_3 plants benefit the most from increased CO_2 levels. Because of their inherently better photosynthetic efficiency, C_4 plants are more efficient users of water, but as Rogers and Dahlman (1990) point out, increased water-use efficiency is an important consequence of increased CO_2 in both C_3 and C_4 plants.

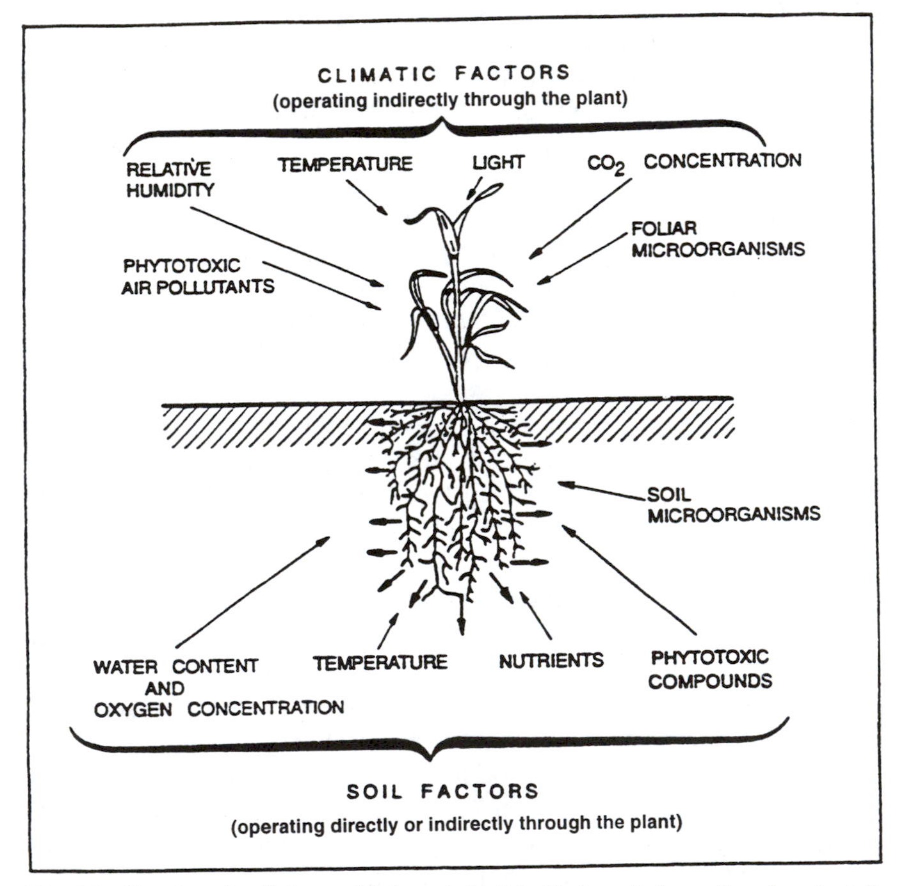

Fig. 7.3. Climatic and soil factors (biotic and abiotic) affecting plant growth and productivity. (Modified from Dommergues, 1978)

Increased CO_2 levels cause shifts in the partitioning of assimilated carbon among different plant parts (e.g., roots, leaves, stems, and pods) (Rogers et al., 1994). Rogers and Dahlman (1990) suggest that a doubling of average CO_2 concentrations could lead to >50% increases in potato and alfalfa yields and almost 30% increases in the yields of corn and soybeans if all other environmental factors remain unchanged. In reality, a crop's health is a product of its interaction with the total environment (Fig. 7.3). Such an interaction is highly random in its nature, exhibiting significant time-dependent dynamics. Such behavior must be coupled with the dynamics of climate (both physical and chemical) change. Because of the spatial variation in the climate and prevalent feedback mechanisms, present predictions of the impacts of climate change on agriculture include significant uncertainties. Because of these considerations, better models are needed to examine the interactive effects of CO_2 with other environmental factors.

One of these is the level of UV-B radiation reaching the earth's surface as a result of stratospheric ozone depletion. Teramura (1990) has stressed the wide range of sensitivities of plants to UV-B: some species are sensitive to current levels, while others are apparently unaffected by massive enhancements. The genetic basis of sensitivity is revealed in comparisons of soybean cultivars: Essex is sensitive, and Williams is tolerant. However, such sensitivity can be masked by other stresses, such as drought. At present, we know very little about the interactions between UV-B and other variables involved in climate change, although such interactions will produce profound effects on the nature of quantitative changes in growth and the quality of crop yield (Krupa and Kickert, 1989).

Another interactive component of climate change is the current global spread of tropospheric O_3 pollution. Available evidence shows that current surface-level O_3 concentrations can result in substantial crop losses in many parts of the world (Elsevier Applied Science, 1988; Heck et al., 1988; Marie and Ormrod, 1990). Other environmental stresses, particularly drought, can dramatically modify the response to O_3; severe drought tends to minimize the adverse effects of O_3. There is also evidence indicating that high CO_2 concentrations and warm air conditions decrease O_3 sensitivity. However, much more needs to be known about these and other interactions before it will be possible to determine the relative importance to crop production of changes in CO_2, UV-B, and O_3 levels, with or without the influence of temperature and water availability.

A Brief Assessment

At the present time, our knowledge of air pollutant effects on crops is substantial compared with what was known in the 1950s and 1960s. We now understand much more about chronic effects as opposed to acute effects, which were emphasized in the past. It is well recognized that growth reduction and productivity losses are critical compared with simple foliar injury, with the exception of crops in which the leaves are the consumed product and ornamental crops that are of aesthetic value. However, our knowledge of the effects of air pollutants on **crop quality** (not quantity) is very limited. Furthermore, much of what we know about air pollution effects on crops is based on studies involving one pollutant at a time. The result is a great deal of uncertainty about the effects of pollutant mixtures (as in the ambient environment) and, in particular, all the variables involved in global climate change (see Chapter 5). Compounding these problems is the interaction between air pollution, microbial pathogens, insect pests, and other environmental factors (Fig. 7.3). While some microbial crop pathogens increase in polluted environments, others decrease. Insect pests frequently increase under O_3 pollution. In one case, the joint effects of O_3 pollution and insect pests resulted in a more rapid senescence of a crop compared with the effects of either stress alone.

Inasmuch as air pollutants can cause direct injury to crop foliage, because of the frequent reduction in the rate of carbon assimilation or photosynthesis, carbon allocation to roots can also be diminished, as in the case of radish (Color Plate 6). Perennial crops such as alfalfa depend on the carbon (starch) reserves in their roots to overwinter. Thus, the effects of air pollutants on roots is as important as those on shoots. This is an area that has not been well researched.

It is not the intent of this assessment to confuse the reader, but to bring attention to the complexities of real world problems. As someone once said, "What we know is only a minuscule amount of what we do not know." It is very important to develop predictions of the impacts of air quality on crop production as working hypotheses. It is equally important to be aware of the associated uncertainties and work toward reducing the consequence of such uncertainties in future crop- and food-production strategies (see Chapter 9, Control Strategies for Air Pollution, and Chapter 10, Education, Research, and Technology Transfer: An International Perspective).

References

Dekker, W. L., and Achununi, V. R. 1990. Greenhouse warming and agriculture. Paper 90-151.2 in: Proc. Annu. Meet. Air Waste Manage. Assoc., 83rd. Air and Waste Management Association, Pittsburgh.

Dommergues, Y. 1978. The plant-microbe system. Pages 1-37 in: Interactions Between Non-pathogenic Soil Microorganisms and Plants. Y. Dommergues and S. V. Krupa, eds. Elsevier Scientific, Amsterdam.

Dreisinger, B. R., and McGovern, P. C. 1970. Monitoring SO_2 and correlating its effects on crops and forestry in the Sudbury area. Pages 12-28 in: Proc. Conf. Impact Air Pollut. Veg. S. N. Linzon, ed. Air Pollution Control Association, Pittsburgh.

Elsevier Applied Science. 1988. Response of crops to air pollutants. Environ. Pollut. 53:1-478.

Guderian, R., and Stratmann, H. 1962. Freilandversuche zur Ermittlung von Schwefeldioxidwirkungen auf die Vegetation. I. Teil: Übersicht zur Versuchsmethodik und Versuchsauswertung, Köln und Opladen. Westdeutscher Verlag, Forsch. Ber. d. Landes Nordrhein-Westfalen Nr. 1118.

Haase, E. F., Morgan, G. W., and Salem, J. A. 1980. Field surveys of sulphur dioxide injury to crops and assessment of economic damage. Paper 80-26.2 in: Proc. Annu. Meet. Air Pollut. Control Assoc., 73rd. Air Pollution Control Association, Pittsburgh.

Heck, W. W., Taylor, O. C., and Tingey, D. T., eds. 1988. Assessment of Crop Loss from Air Pollutants. Elsevier Applied Science, London.

Hidy, G. M., Mahoney, J. R., and Goldsmith, B. J. 1978. International Aspects of the Long-Range Transport of Air Pollutants. Final Report. U.S. Department of State, Washington, DC.

Jacobson, J. S., and Showman, R. E. 1984. Field surveys of vegetation during a period of rising electric power generation in the Ohio River Valley. J. Air Pollut. Control Assoc. 34:48-51.

Kohut, R. J., Amundson, R. G., Laurence, J. A., Colavito, L., van Leuken, P., and King, P. 1987. Effects of ozone and sulfur dioxide on yield of winter wheat. Phytopathology 77:71-74.

Krupa, S. V., and Kickert, R. N. 1989. The greenhouse effect: Impacts of ultraviolet (UV)-B radiation, carbon dioxide (CO_2) and ozone (O_3) on vegetation. Environ. Pollut. 61:263-392.

Krupa, S. V., and Manning, W. J. 1988. Atmospheric ozone: Formation and effects on vegetation. Environ. Pollut. 50:101-137.

Krupa, S. V., Tonneijck, A. E. G., and Manning, W. J. Ozone. In: Recognition of Air Pollution Injury to Vegetation: A Pictorial Atlas. R. B. Flagler, A. H. Chappelka, W. J. Manning, P. M. McCool, and S. R. Shafer, eds. Air and Waste Management Association, Pittsburgh. (In press.)

Manning, W. J., and Feder, W. A. 1980. Biomonitoring Air Pollutants with Plants. Applied Science, London.

Manning, W. J., and Krupa, S. V. 1992. Experimental methodology for studying the effects of ozone on crops and trees. Pages 93-156 in: Surface Level Ozone Exposures and Their Effects on Vegetation. A. S. Lefohn, ed. Lewis, Chelsea, MI.

Marie, B. H., and Ormrod, D. P. 1990. Effects of tropospheric ozone on plants in the context of climate change. Paper 90-151.3 in: Proc. Annu. Meet. Air Waste Manage. Assoc., 83rd. Air and Waste Management Association, Pittsburgh.

McLaughlin, S. B., and Lee, N. T. 1974. Botanical studies in the vicinity of Widow's Creek Steam Plant. Review of Air Pollution Effects Studies, 1952-72 and Results of 1973 Surveys. Intern. Rep. I-EB-74-1. Tennessee Valley Authority, Muscle Shoals, AL.

Rogers, H. H., and Dahlman, R. C. 1990. Influence of more CO_2 on crops. Paper 90-151.1. in: Proc. Annu. Meet. Air Waste Manage. Assoc., 83rd. Air and Waste Management Association, Pittsburgh.

Rogers, H. H., Runion, G. B., and Krupa, S. V. 1994. Plant responses to atmospheric CO_2 enrichment, with emphasis on roots and the rhizosphere. Environ. Pollut. 83:155-189.

Runeckles, V. C., and Krupa, S. V. 1994. The impact of UV-B radiation and ozone on terrestrial vegetation. Environ. Pollut. 83:191-213.

Teramura, A. H. 1990. Ozone depletion, ultraviolet light and plants. Paper 90-151.4 in: Proc. Annu. Meet. Air Waste Manage. Assoc., 83rd. Air and Waste Management Association, Pittsburgh.

Treshow, M. 1984. Introduction. Pages 1-6 in: Air Pollution and Plant Life. M. Treshow, ed. John Wiley & Sons, New York.

Further Reading

Guderian, R. 1977. Air Pollution. Phytotoxicity of Acidic Gases and Its Significance in Air Pollution Control. Ecol. Stud. 22. Springer-Verlag, New York.

Guderian, R., ed. 1985. Air Pollution by Photochemical Oxidants: Formation, Transport, Control and Effects on Plants. Springer-Verlag, New York.

Treshow, M., and Anderson, F. K. 1989. Plant Stress from Air Pollution. John Wiley & Sons, New York.

Wellburn, A. 1994. Air Pollution and Climate Change: The Biological Impact. John Wiley & Sons, New York.

CHAPTER 8

Air Quality and Forests

> Since 1982 and after some 600 research projects at a cost of 275 million DM (approximately $170 million), scientists so far cannot clarify the reason for tree decline.
>
> —*Schwäbische Zeitung,* October 4, 1989

Introduction

The mythical bonds between man and the forest go back to ancient times (Prinz, 1987). Thus, the problem of adverse human influence on forest lands has its roots far in the past. Large settlements in the temperate regions of the world were often established in or near hardwood or conifer forests in order to have easy access to basic necessities, such as fuel, building materials, game, and a steady supply of pure water (Miller and McBride, 1975). In the beginning, these forests conformed to the standard defined as "a plant association predominantly of trees or other woody vegetation occupying an extensive area of land" (Committee on Forest Terminology, 1944). Tree density was sufficient enough over large areas so that distinct climatological and ecological conditions developed that were easily distinguishable from those of less densely vegetated areas. The forest ecosystem, with its distinct association of plants, animals, and many other organisms (both macroscopic and microscopic) organized to utilize and transfer energy and raw materials, functioned efficiently under the influence of its controlling physical environment (Billings, 1970).

The typical forest in a populated region of the world today is not usually a broad expanse of green woodland stretching to the horizon. More often, we see remnants of a forest that has been modified in many ways to suit human purposes (e.g., the deforestation of the Amazon).

These forest remnants may no longer operate with the natural efficiency of the undisturbed ecosystem from which they were derived.

Frequently, heavy industry has been located in forested areas because of the coincident availability of rich ore deposits. The most damaging to forests have been nonferrous metal smelters, aluminum ore reduction plants, and coal-burning power plants (Table 8.1). The transport of oxidant (e.g., O_3) air pollutants from urban centers to forests up to 100 miles (160 km) downwind illustrates the fact that a long distance between the sources and the forest offers no protection from severe injury (Color Plate 7) (Miller and McBride, 1975).

Chapter 7 provides additional summaries of acute foliar injury symptoms on trees caused by individual air pollutants (see Tables 7.2, 7.3, and 7.5) and sensitivity of tree species to some of those air pollutants (see Table 7.6).

An interesting example of the effects of chronic air pollutant (e.g., SO_2) exposure on tree growth and phenology was provided by A. H. Legge (*personal communication;* Color Plate 8). Chronic exposure to air pollutants such as O_3 and SO_2 year after year can delay the spring onset of tree growth (Color Plate 8) and also initiate premature senescence and defoliation in the fall (Fig. 8.1). Both of these processes result in a com-

Table 8.1. Some well-known examples of forest damage caused by sulfur dioxide (SO_2) and fluoride

Pollutant and location	Pollutant source	Year
SO_2		
Western U.S. and Canada		
Anaconda, MT	Copper smelter	1906–1907
Missoula, MT	Pulp and paper mill	1963
Redding, CA	Copper smelter	1903–1904
Trail, British Columbia	Copper smelter	Early 1900s
Eastern U.S. and Canada		
Copper Basin, TN	Copper smelter	1908
Cumberland Plateau, TN	Pulp mill, coal-burning power plant, uranium-refining mill, ferro-alloy reduction plant	1964
Sudbury, Ontario	Nickel and copper smelter	Early 1900s
Wawa, Ontario	Iron sintering plant	1960s
Europe		
Locations in Czechoslovakia, Poland, and Scotland	Number of sources	1950–1960
Ruhr Valley, Germany	Number of sources	1960s
Fluoride		
Western U.S.		
Columbia Falls, MT	Aluminum ore reduction plant	1955
Spokane-Mead, WA	Aluminum ore reduction plant	1943
Europe		
Norway	Aluminum plant	1960s
Raushofen, Austria	Aluminum plant	1960s

pressed annual growth period and finally loss of tree vigor and decline. These effects are clearly seen in the polluted area of southern Poland as well as in many other locations. In conifers, these effects are preceded by a lack of retention of older needles in the stressed trees.

History of Forest Damage

With the acceleration of industrial expansion during the 19th century, damage to vegetation became more readily apparent. In the worst cases, this resulted in the appearance of plant "deserts" surrounding highly industrialized British cities such as Manchester and Liverpool. In Germany toward the end of the 19th century, industrial releases of extremely high levels of air pollutants in the valleys of the Erzgebirge caused

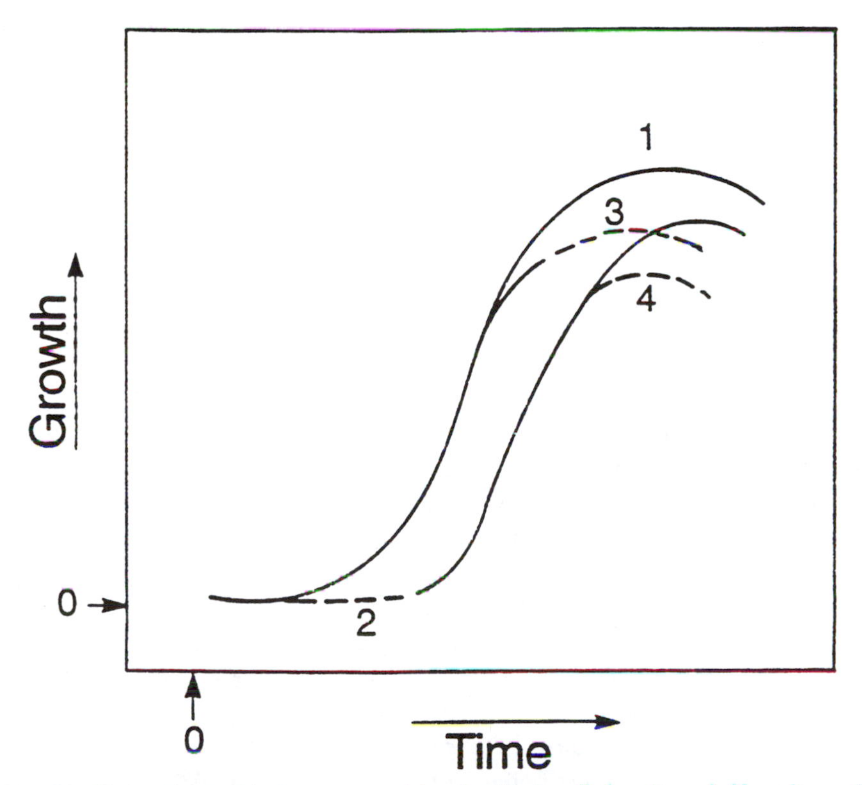

Fig. 8.1. Changes in growth patterns caused by chronic air pollution stress. **1,** Normal annual growth curve. **2,** Altered growth pattern resulting from delay in the onset of early growth. **3,** Altered growth curve resulting from premature senescence or early termination of growth in the fall. **4,** Altered growth curve resulting from both delay in the onset of early growth and early termination of growth in the fall.

extensive forest devastation and stimulated much early research, especially by the forestry faculties at the Universities of Tharandt and Freiberg in the former East Germany. The impressive state of knowledge at that time was summarized by Wislicenus (1907).

According to Wislicenus, acute injury could be distinguished from chronic (or "hidden") injury. Sulfur dioxide was the main cause of damage in both cases, but other "classic" air pollutants were also involved. Conifers proved to be more sensitive to long-lasting exposure to low pollutant concentrations (chronic effects), while deciduous trees were more sensitive to short-term high concentrations (acute effects). There was a close correlation between metabolic activity and sensitivity. Leaves and needles were most sensitive when just reaching full development, and whole plants suffered much more when fumigated during the day than during the night with the same concentration. Conifers also proved to be much more sensitive during summer than winter months. Prevailing winds determined the location and shape of the damaged area. On the basis of wind direction, clear shadow effects were observed within stands as well as within the crowns of single trees. Poor soil nutrition enhanced the effect of air pollution; however, more important to damage severity was the soil water content, particularly since drought reduced susceptibility to injury.

This classic type of forest decline was named *Rauchschäden* (**smoke damage**). Scientific proof of its causes was established through observations of the spatial and temporal relations between intensity of damage and the existence of specific sources of air pollution. A number of insightful experiments were also conducted at that time. Clear proof concerning chronic effects was of course lacking, so the hypotheses about this type of damage had to be inferred from observations in the high range of air pollution concentrations causing acute effects.

As early as 1911, there were attempts to replace the *Rauchschäden* theory with the assumption that acute damage could appear only if previous acidification of soil by deposition of sulfuric acid (H_2SO_4) had taken place (Wieler, 1911). The eventual rebuttal of this view was based on the observation that damage also occurred on calcareous soils and that this damage quickly disappeared when the sources of air pollution were controlled or were no longer in operation.

Another striking finding of early studies was that not all kinds of forest damage can be attributed simply to anthropogenic (human) environmental influences. For example, in the first decades of this century, extensive damage to silver fir, a phenomenon called *Tannensterben* (**fir decline**), occurred at the northern border of this tree's natural habitat.

The causes of the *Tannensterben* were never unequivocally identified, but air pollution as a causal agent was never really discussed (Neger, 1908). Compared with the situation of today, these early episodes of forest damage were neither systematic and widespread nor intensive. With good reason, then, the more recent phenomenon of extensive forest damage in Germany was termed *neuartige Waldschäden* **(novel forest damage)**.

As a general conclusion, it can be stated that for the classic type of forest decline, especially in its chronic form, only circumstantial evidence of causation by air pollutants was provided. Nevertheless, the influence of air pollutants was never doubted where their concentrations in space and time were in agreement with the pattern of damage. Much later, in the Biersdorf (Germany) field study during the 1960s (Guderian and Stratmann, 1968), the first relevant quantitative relationships on a really broad scale were produced. Disadvantageous impacts on trees resulting in growth reduction started with SO_2 levels of roughly 50 $\mu g\ m^{-3}$ (approximately 20 ppb) as an annual arithmetic mean. The sigmoidal (∫) shape of the dose-effect relationship was most striking: low dosages caused relatively little damage until a threshold was exceeded. Damage then increased rapidly with further dosage increases until a dose was reached at which essentially death occurred.

Recent Forest Damage

Any serious attempt to clarify the nature and causes of forest damage must address the following issues:

- Different symptoms on different tree species (and even on the same tree species) have to be clearly identified (Table 8.2) and precisely described before their causes are investigated.
- It is necessary to distinguish between long-term trends and short-term fluctuations in the development of damage. Although air pollution is an example of a causal factor in long-term trends, an air pollution episode triggered by meteorological events may be a cause of short-term fluctuations. In each case, however, valid statements about trends can be derived only from appropriate time-series analyses and not from a few single observations.
- Likewise, with respect to the spatial distribution of damage, broad-scale variation has to be distinguished from fine-scale variation.

Table 8.2. Common symptoms of tree decline (*Waldsterben*) in central Europe, 1985[a]

Symptoms of growth reduction	Symptoms of abnormal growth
Discoloration and loss of foliar biomass	Active casting (abscission) of green leaves and needles
Loss of feeder root biomass[b]	Stork's nest formation in white fir[c]
Decreased annual growth increment	Altered branching habit and greater than normal production of adventitious shoots[d]
Premature senescence of older needles in conifers	Altered morphology of leaves
Increased susceptibility to secondary root and foliar pathogens	Altered allocation of carbon (biomass)
Death of affected trees	Excessive seed and cone production
Death of herbaceous understory vegetation	

[a] Modified from Schütt and Cowling, 1985.
[b] **Feeder roots** = secondary, fine roots on a tree that are essential for water and nutrient absorption. Unlike primary roots, these roots are not woody and are not involved in anchoring the tree.
[c] **Stork's nest** = shape similar to that of a nest built by a stork on top of a tree to lay eggs.
[d] **Adventitious shoots** = shoots formed on the trunk of a tree in abnormal places; i.e., not normal branches.

• A complex causal scheme results when it is recognized that air pollutants can attack plants directly as well as indirectly via the soil.

In summary, it can be concluded that the cause of the recent type of forest damage reported from Europe and North America is a complex interaction between different anthropogenic (human) and nonanthropogenic (natural) factors. In addition, the recent type of forest decline reflects a chronic process, which is associated with internal dynamics as well as the accumulation of toxic materials coupled with other complicating factors (Fig. 8.2). This is again a distinction from the classic smoke damage, where acute effects were clearly predominant.

Damage in North America

In 1985, a review of the facts about forest damage in North America showed that unlike the situation in the former West Germany, precise data on the development and distribution of forest damage were missing as were data on air pollution in the most affected areas (Prinz, 1985). Since then, impressive efforts to investigate the causes of forest decline have been undertaken. Most of these investigations were part of the Forest Response Program sponsored by the National Acid Precipitation Assessment Program (NAPAP) and were funded by the U.S. Environmental Protection Agency (EPA), the U.S. Forest Service (USFS), and the National Council of the Paper Industry for Air and Stream Improvement (NCASI).

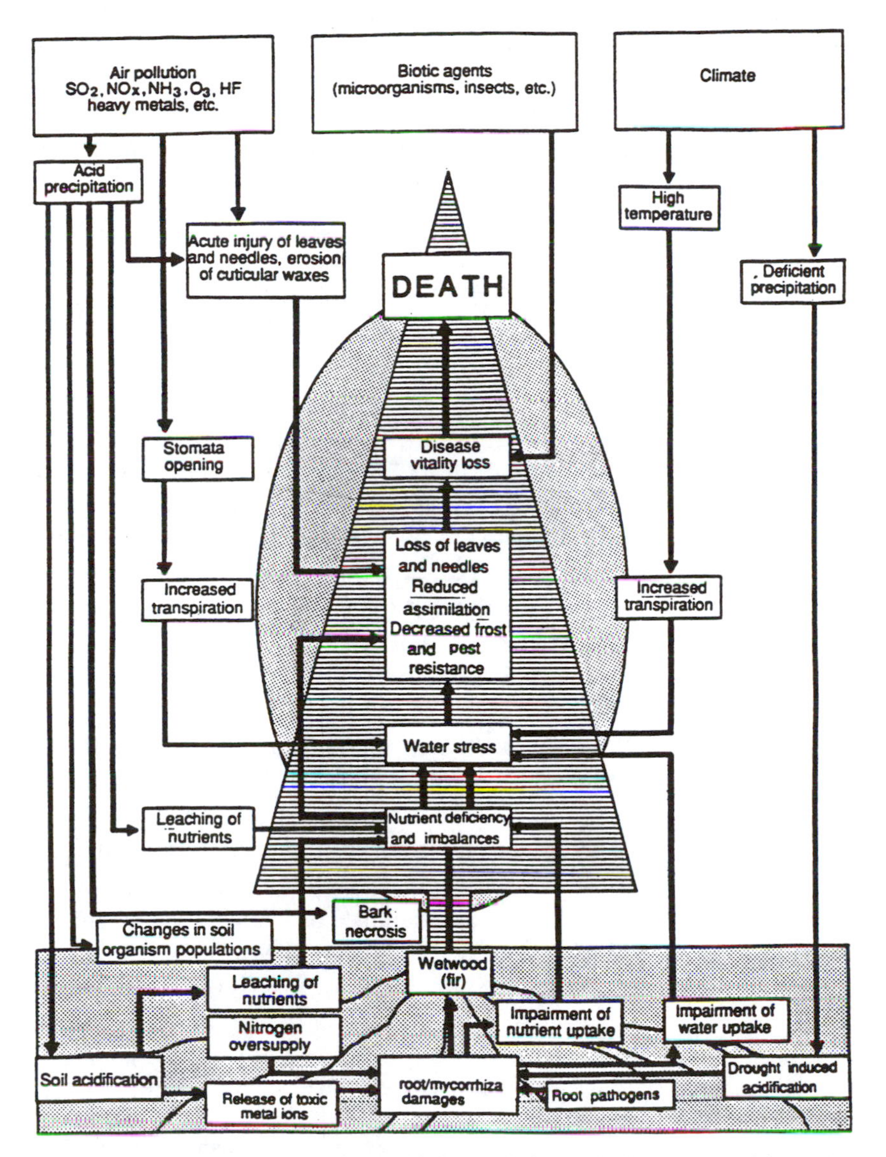

Fig. 8.2. Various interactive factors that could play a role in tree decline. (Reprinted from Schäfer et al., 1988, by permission of Kluwer Academic Publishers)

The most severe tree decline in North America was seen in red spruce (*Picea rubens*) growing at high elevations from New York and New England to the southern Appalachian Mountains. Tree-ring analyses indicate a substantial decrease in tree growth rate since about 1970 as well as a strong association between severity of damage and elevation, especially in the northern forests (McLaughlin, 1985). Both features are comparable to the situation in Germany. In southern Germany, for example, the first symptoms of growth reductions were traced back to the early 1960s (Kenk et al., 1984).

The symptoms of forest damage are rather diverse, especially if other tree species such as balsam fir (*Abies balsamea*) and Fraser fir (*A. fraseri*) are included. In North America, a major visible foliar symptom is the loss of needles starting at the tips of branches and at the apex of the crown, but unlike the situation in West Germany, needle loss is not accompanied by pronounced chlorosis or symptoms of nutrient deficiency.

The ideas of the German scientist Bernhard Ulrich (1984) were adopted early in the 1980s, and the forest decline in North America was attributed to soil acidification or aluminum toxicity resulting from acid deposition (Wetstone, 1983). These assumptions were soon challenged, however, because the mechanisms of aluminum toxicity are inhibited by organic matter in the soil and organic matter is prevalent at the high elevations where forest damage is most intense. Furthermore, except for the most sensitive soils, leaching of nutrients did not appear to pose a threat to soil base-cation supplies in the United States (Johnson et al., 1985).

Other researchers suggested that the abnormally high frequency of drought periods during the 1960s could have caused the forest damage or at least could have triggered it (Johnson, 1986). More recently, the impact of ozone as a major causal factor has gained much attention. Another recently discussed suggestion is that enhanced input of nitrates (NO_3^-) from the past may have weakened tree resistance to frost and that winter damage to foliage may be the crucial factor in forest decline (Friedland et al., 1984). McLaughlin (1985) concluded that

> . . . based on field irrigation experiments and studies of forest nutrient cycles, short-term negative effects of acid deposition on forest soils appear unlikely. Longer-term potential for negative effects . . . has not yet been adequately evaluated. . . . There are still many postulated pathways and mechanisms for the observed responses, including both direct and indirect influences of O_3, wet and dry deposited strong acids, and heavy metals, as well as climatic change. . . . However, there is still no clear evidence of a single causal agent.

It should be added, however, that the encouraging and promising research underway in the United States may provide a much better picture than the one available in the past. It seems that these investigations, in combination with other factors, will play a significant role.

Damage in Europe

Eastern Europe

In some subalpine areas of Poland, Czechoslovakia, and the former East Germany (especially the Erzgebirge), exist the worst cases of forest damage in the world. This damage has been caused solely by sulfur dioxide and other primary air pollutants and therefore has to be associated with the classic type of forest decline. A recent report, however, shows that in the cleanest areas of Czechoslovakia (with respect to sulfur dioxide concentration), dramatic forest damage, especially in Norway spruce stands, took place during 1982 and 1983 (Materna, 1985). There are no obvious causes for this damage: the concentration of sulfur dioxide was definitely below the threshold levels of 40–50 $\mu g\ m^{-3}$ (16–20 ppb) identified in the Biersdorf (Germany) field trial. The clear deficiencies of magnesium and potassium in the fir and spruce populations resemble the symptoms found in West German alpine and subalpine forests.

Widespread forest damage (in 1985 about 70,000 hectares) with unclear causality has also been reported recently from Slovenia (Solar, 1986). According to needle analysis, "dry sulfur deposition, except in the surroundings of known local pollution sources, cannot be a significant reason" for this phenomenon. Many symptoms were similar to those of the *neuartige Waldschäden* **(novel forest damage)** in the former West Germany.

Nordic Countries

Inventories of forest damage in Sweden showed that spruce trees were suffering from needle loss primarily in the most southern and western parts of southern Sweden, especially in dry soil on exposed sites. Older trees were more damaged than younger trees. Early reports speak of a "remarkable coincidence between the map of tree damage and maps showing the areas of lake acidification" (Svensson, 1984–1985). However, 5-year experiments with simulated artificial rain and results from other studies conducted in Norway suggest that acid deposition is unlikely to have rapid or major effects on soil pH in the Nordic countries (Abrahamsen and Stuanes, 1986). Neither needle chlorosis nor apparent nutrient deficiencies occurred in damaged stands of Norway spruce in southern Sweden. A slight positive correlation exists between degree of needle loss and aluminum concentration in the needles and could indeed point to the toxic influence of this element, but the content of aluminum, calcium, magnesium, manganese, iron, zinc, and sodium in the needles showed the same unexpected tendency, that is, positive correlation with needle loss.

A more recent publication concludes cautiously that "no one factor can be singled out as the cause of damage" but that there are many coacting stress factors, such as drought, frost, wind, airborne pollutants (such as ozone, nitrogen oxides, and ammonia), nutrient deficiency, and metal poisoning caused by soil acidification, as well as nitrogen saturation (National Swedish Environmental Protection Board, *unpublished*).

Impacts of Air Pollutants on Forest Ecosystems: An Integrated Analysis with Emphasis on O_3

Kickert and Krupa (1990) published a detailed analysis of the impacts of air pollutants on forest ecosystems. They cited an analysis by Hidy et al. (1978) that projected future energy development country by country around the globe through the year 2025. These spatial patterns of future energy development were evaluated in terms of the global spatial patterns of solar radiation to identify regions on the earth that were likely to be areas of high photochemical smog by the year 2025 (Hidy et al., 1978). Kickert and Krupa (1990) have reproduced a global map of these regions showing which forest types within these regions could be potentially affected in some way by tropospheric O_3 in the future (Fig. 8.3). The forest and woodland types that were used by Olson (1970) in mapping global ecosystem patterns prior to the Iron Age and that are also found in each of the regions likely to be highly susceptible to photochemical smog by the year 2025 include the western and southeastern portions of the United States. Three additional smaller smog regions in North America not found in the original analysis of future photochemical smog regions of Hidy et al. (1978) are shown in Figure 8.3: the Mexico City area in Mexico and the Vancouver-Fraser Valley, British Columbia, and Toronto, Ontario, areas in Canada.

In South America, forest lands in eastern Brazil from Salvador to São Luis are found in one smog region, while forests in other parts of Brazil and in the south central and southern parts of the continent (Chile, Argentina, and Paraguay) are found in a second smog region. On the eastern side of the Atlantic Ocean, forest and woodlands in two regions of high future photochemical smog are found in what is roughly called the "Mediterranean Rim" and in a smaller region in South Africa. The former actually includes portions of countries in southern Europe, northern Africa, and the Middle East. Another region of projected high

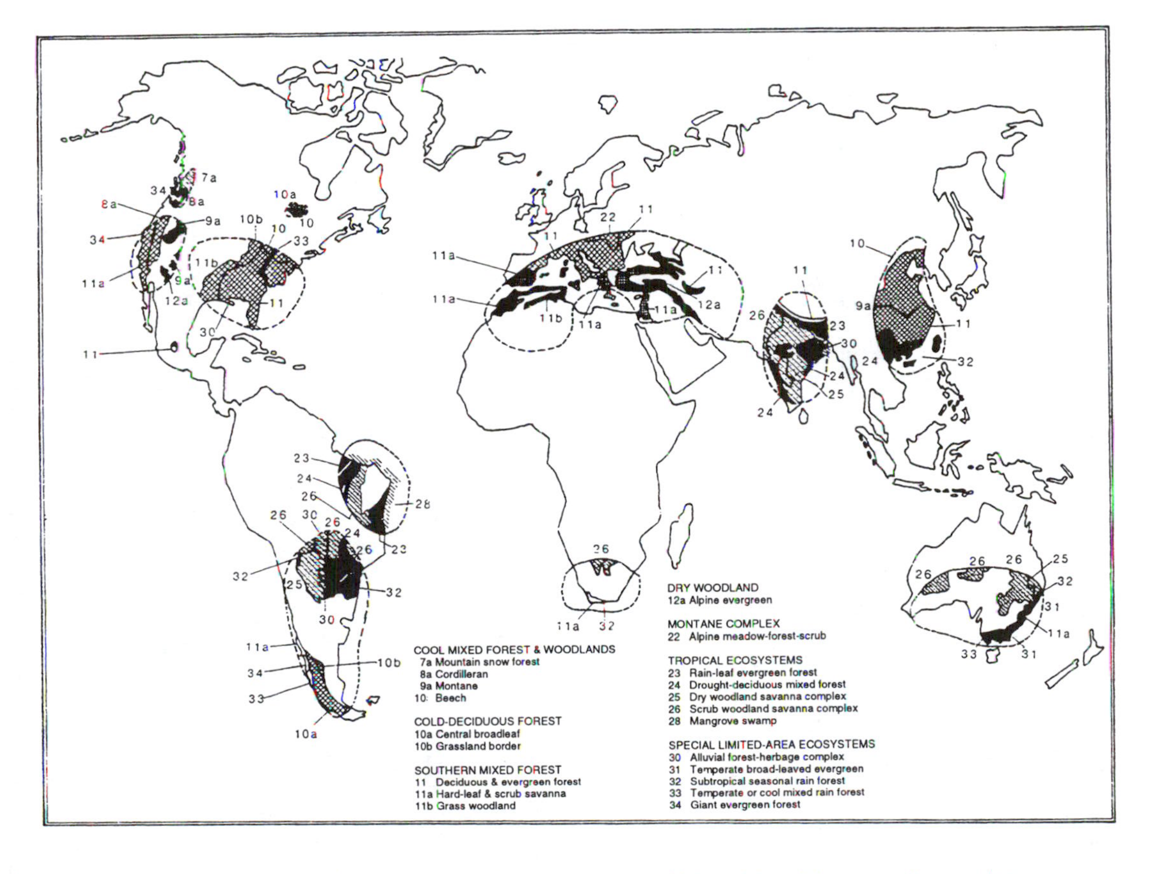

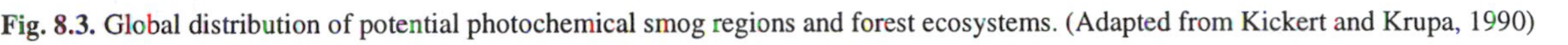
Fig. 8.3. Global distribution of potential photochemical smog regions and forest ecosystems. (Adapted from Kickert and Krupa, 1990)

tropospheric O_3 is a major part of the Indian subcontinent. Two other regions containing vast forest lands where tropospheric O_3 might reach relatively high levels are found in southeastern Asia and southern Australia.

Effects on Physiological Processes of Trees in Artificial Exposures

Since the 1980s, many reviews have been published on the effects of O_3 on forests in North America and western Europe (Krupa and Kickert, 1989). Most of the research has been done on trees rather than on other ecosystem components. Most of these studies have dealt with conifers, although some broad-leaved deciduous hardwoods have been examined. A small number of studies have examined mature trees in forests under ambient O_3 exposure, but much of what we know is based on the exposure of seedlings under artificial conditions, often as monocultures in containers.

Forest and Tree Responses to O_3 Under Field Conditions

Probably the most comprehensive data collection and analysis performed for a forest under ambient conditions is a study conducted during 1972–1980 on the San Bernardino mixed conifer forest in southern California (Miller et al., 1982). While average 24-hr O_3 concentrations during the summer months reached 0.10–0.12 ppm, needle injury could be detected after a 24-hr exposure to O_3 concentrations of 0.05–0.06 ppm. Another forest research program is in progress in North Rhine-Westfalia in Germany (Koth and Prinz, 1988). In the Eggegebirge and Eifel Mountains, annual peak O_3 values were found to be about 0.13 ppm and summertime monthly average values were nearly 0.05 ppm. Unlike the California study, where pines were dominant and O_3 sensitive, here spruce and fir are the dominant species.

Ozone and Interspecies and Intraspecies Sensitivity and Competition for Growth Resources

As with crops, there are differences among tree species in their sensitivity and response to O_3 (Krupa and Kickert, 1989). Most often, these differences have been demonstrated in North America and western Europe by screening seedlings in artificial exposures. Such differences may or may not be evident in established forests with mature trees. It is important to note that screening experiments involve short-term studies, while natural forests are subjected to chronic exposures ranging from decades to >100 years under differing exposure conditions. In addition, there are genotypic differences in sensitivity to O_3. However, it is not possible to extend the sensitivity rankings of North American and European species to other unscreened tree species found around the world in the forest and woodland types shown in Color Plate 7.

No observational or experimental studies have been conducted to examine shifts in competitive balance within a monoculture of trees at different densities or between two or more plant species competing with each other in a mixed species forest community in response to various O_3 exposure patterns and environmental conditions (Guderian and Kueppers, 1980; Runeckles, 1986).

Responses of Forest and Woodland Ecosystem Components and Processes

Energy Flows Through Ecosystems

While data from analyses of energy flow through ecosystems, undisturbed by environmental pollutants, have been published during the 1980s and 1990s, there do not appear to have been any such forest ecosystems analyses performed explicitly for O_3 exposure. Two considerations probably have inhibited such an analysis: 1) forests are not generally managed on the basis of energetics; and 2) ecosystem components that directly represent a "small" pathway for energy flow, for example, invertebrate consumers such as bark beetles, can exert great influences on energy flow in the rest of the system.

Litter Dynamics and Nutrient Cycling

Vegetation as primary producers in forest and woodland ecosystems require nutrients to grow. Just how forest nutrient cycles respond to O_3 concentrations elevated above natural levels is to a large extent an artificial question. In nature, many forests will also be subjected to inputs of SO_2 and/or other gases, dry or wet deposition of other sulfur and nitrogen compounds, and additions of many anions and cations. Rates of nutrient cycling will vary from one forest to another because of differences in soil fertility and the nutritional requirements of the various plant species and their abundance. In terms of O_3 and forest nutrient cycling, there is but one relatively old review on the subject (Zinke, 1980).

As they do in crops (Table 7.8), increases in tropospheric O_3 can reduce photosynthesis and dry matter production in some O_3-sensitive tree species (Krupa and Kickert, 1989). This amounts to a decrease in the rate of carbon cycling in a forest ecosystem containing these species and has two implications: 1) it is unclear whether, under sequential or simultaneous exposure to elevated levels of O_3 and CO_2, O_3-sensitive tree species would have any effect on the rate of carbon cycling; and 2) aside from changes in ambient CO_2 levels, a decrease in the rates of growth and carbon cycling in O_3-sensitive tree species could decrease the rate of cycling of other nutrients.

In big-cone spruce (*Pseudotsuga macrocarpa*) in the mountains of southern California, the live foliage N-P (nitrogen to phosphorus) ratio increased along spatial gradients of increasing oxidant pollution (Zinke, 1980). The cycling of N was accelerated while the cycling of P was retarded. The O_3-sensitivity status of big-cone spruce seedlings with respect to biomass growth (carbon flow) appeared to resist dry weight change under O_3 exposure.

Leaf-Fall, Litter Accumulation, and Decomposition

In the San Bernardino Mountains, average annual needle litter-fall of ponderosa (*Pinus ponderosa*) and Jeffrey (*P. jeffreyi*) pines increased by as much as 270% (2.7 times) as oxidant-induced foliar injury increased (Arkley and Glauser, 1980). Needle abscission rates were directly proportional to needle age and average annual O_3 dose (Miller and van Doren, 1982). The percentage of needle nutrient content of P, N, and K (potassium) increased, and Ca (calcium) decreased in the litter-fall as the degree of injury to the living foliage increased (Arkley and Glauser, 1980; Fenn and Dunn, 1989). In contrast, lower amounts of P, N, K, Ca, and Mg (magnesium) were found in live foliage of ponderosa pine, indicating that the processes of nutrient conservation within these trees prior to needle fall had been disrupted (Miller et al., 1982). In this forest, surface needle litter decomposed faster under greater oxidant pollution than in less polluted geographic areas (Fenn and Dunn, 1989).

Foliage and Root Storage of Nutrients

Controlled chamber fumigation of potted eastern white pine (*P. strobus*) seedlings with O_3 concentrations of up to 0.14 ppm for 1.25 hr per day, 3 days per week, alternating with simulated acidic rain ranging from pH 5.6 to 3.0 at 12.5 mm per day, 2 days per week for 15 weeks, showed no effect of O_3 on root and foliage content of N, P, Mg, Al (aluminum), and Fe (iron) or on nutrient weight ratios. Effects of O_3 were observed on the foliar concentrations of K, Ca, Mn (manganese), and Cu (copper) and on root Ca and Mn.

Leaching of Nutrients from Foliage

One explanation of the multiple chain of events leading to the novel forest decline in the former West Germany involves the effects of ambient O_3 on the nutrient cycles of Norway spruce (*Picea abies*) (Krause and Prinz, 1989). Some investigators believe that exposure to elevated levels of O_3 enable accelerated leaching of Mg, Ca, Mn, and Zn (zinc) from foliage by acidic precipitation of one form or another (Krause and Prinz, 1989). The resulting tree foliar nutrient deficiency leads to root injury and decreases nutrient uptake from the soil, thereby setting off a positive feedback that amplifies tree nutrient deficiency in the ecosystem. However, some laboratory experiments conducted over a 14-month period found nutrient leaching to be negligible. Other investigators reported that the chlorotic needle symptoms in Norway spruce and silver fir (*Abies alba*) are typical of Mg deficiency, and the effects of air pollution (O_3, SO_2, or acid mist) on foliar leaching could not be reproduced in the laboratory in short-term experiments to the extent needed to reproduce these symptoms (Roberts et al., 1989). According to these authors, foliar symptoms are related only to Mg-deficient soils and to forest management practices (harvesting). According to still another report, field measurements do not show long-term harmful effects from gaseous pollutants implicated in Norway spruce needle yellowing, growth decline, or mineral cycles in southern Germany (Schulze, 1989).

In an elaborate experiment conducted over a 5-year period, 5- to 8-year-old Norway spruce, silver fir, and European beech seedlings along with ground vegetation, litter, humus, and transplanted forest soil were artificially exposed to O_3 at concentrations of 0.04–0.06 ppm, 7 hr per day, for the first half of the experiment; 0.015–0.092 ppm, 12 hr per day, for the remaining period; and simulated acidic precipitation (pH 4.0) (Seufert et al., 1989). Exposure to O_3 alone (no SO_2 added) increased Zn in the throughfall. The other anions and cations (SO_4^{2-}, Mg^{2+}, Ca^{2+}, Mn^{2+}, and soil Al^{3+}) showed little response to O_3 alone. In treatments with a mixture of O_3 and SO_2, it appeared that O_3 acted as a catalyst, compared with SO_2 alone, in enhancing the leaching and throughfall of Mn^{2+} and H^+.

Nutrient Condition of Trees and Their Response to O_3

Just as O_3 can affect forest mineral cycles, there is potential for a positive systems-level feedback acceleration of the degradation of a forest through a decrease in nutrients that might influence tree responses to O_3. In Norway spruce forests that are not fertilized, one might suspect that even though low nutrient supply might not affect visible injury response to O_3, nutrient stress could accelerate the decrease in the photosynthetic capacity and in addition reduce frost resistance in older (not current-year) foliage (Höpker et al., 1989).

Moisture Flows Through Ecosystems

Watershed hydrologic responses as a consequence of chronic effects of O_3 on vegetation are probably more significant in conifer forests and in humid temperate zones, although few observations and analyses have been published. Results from studies on the spruce forests in the Ore Mountains along the border of the former Czechoslovakia and East Germany that are exposed to heavy industrial air pollution showed a decrease in canopy interception with defoliation and increases in the rate of snow melt, soil erosion, soil moisture in the root zone, and runoff (to streamflow) (Materna, 1984). Tree injury attributed to high deposition of acidic pollutants in an uninhabited, uncut Norway spruce watershed of the northern Black Forest in the former West Germany has been attributed to significant increases in water yield during 1973–1986 (Caspary, 1989).

The mixed conifer forest of the San Bernardino Mountains of California is an area of relatively low precipitation perched above surrounding arid lands with a Mediterranean summer–dry climate. The forest's annual average summer water use over 5 years during the mid-1970s was compared with average foliar O_3 injury for ponderosa and Jeffrey pines (Kickert et al., 1980; Arkley, 1981). There was some indication that as foliar O_3 injury increased, summer soil water use declined, allowing soil moisture to increase. However, in this ecosystem, trees were extracting moisture at great depths (>9 m) in the decomposed granite underlying the soil (Arkley, 1981). This usually mitigates any intermittent increase in excess soil moisture diverted to streamflow.

In humid conifer forests dominated by one or two species of mature trees, bark beetle epidemics might give an indication of what could happen to streamflow after several years of chronic O_3 exposure. A bark beetle epidemic occurred during 1939–1945 near the Continental Divide in Colorado in two Engelmann spruce (*Picea engelmannii*) and subalpine fir (*Abies lasiocarpa*) watersheds (Bethlahmy, 1975). From 1946 to 1960, the average annual streamflow increased significantly in both watersheds, and even 25 years later, water yields were still greater than prior to the epidemic (Bethlahmy, 1974).

Microbial Consumers and Diseases of Forest Primary Producers

We know very little about how the effects of microbial consumers and their associated tree diseases change in forests subjected to chronic O_3 exposure. The most comprehensive review paper on the subject is now more than a decade old (Treshow, 1980). There is ample evidence to suggest that the roots of stressed trees exhibit inadequate and declining beneficial, symbiotic relationships with soil microorganisms (e.g., mycorrhizae), a relationship that is obligatory for tree health and vigor. Those few biotic diseases that have been studied relate to North American tree species. The only general principle identified is that fungal pathogens that require weakened host trees will show either no effect or increased activity under chronic exposure to air pollution, while populations of others attacking healthy host trees will decline (Treshow, 1980; Coleman et al., 1987).

Invertebrate Consumers

Although there are review papers on the effects of air pollution on insects (e.g., Führer, 1985), we know relatively little about the response of mature tree stem- and leaf-infesting invertebrates in forest ecosystems under ambient O_3 exposures. In conifer forests in the western United States, it takes fewer western pine bark beetles (*Dendroctonus brevicomis*) to kill ponderosa pines whose needles are shorter than normal because of chronic ambient oxidant exposure than to kill healthy trees (Dahlsten et al., in press). Such trees also are more active in producing early summer broods of beetles compared with healthy trees. It was also found that a given number of western pine bark beetles can kill more oxidant-injured trees than healthy trees. From an historical analysis of insect outbreaks during the past 150 years in European conifer forests of the Mediterranean Rim, Baltensweiler (1985) found that five leaf-infesting insects that were hardly known had become new pests on Norway spruce since about 1940. This is only a circumstantial observation related to increasing levels of air pollution, and because of the time lags and complex interactions in forest ecosystems, the determination of such cause-and-effect relationships might not be the most appropriate.

In experiments with 0.10 ppm O_3 for 7 hr per day, 5 days per week for 9 weeks, in temperate, cold deciduous forests, elm trees (*Ulmus* sp.) appeared to suffer enhanced foliar herbivory because the number of eggs per female elm leaf beetle (*Xanthogaleruca luteola*) was found to increase compared with the clean-air treatment (Hall et al., 1988). In greenhouse experiments in which stem cuttings of eastern cottonwood (*Populus deltoides*) were exposed to O_3 at 0.20 ppm for 5 hr, no effect was found on the ant-tended aphid (*Chaitophorus populicola*), but its competing pests, the leaf rust fungus (*Melampsora medusae*) and the willow leaf beetle (*Plagiodera versicolora*), showed reduced reproduction (Coleman and Jones, 1988). Consequently, lowered competition under O_3 exposure might allow greater aphid injury on eastern cottonwood. On white oak (*Quercus alba*), the seedling foliage was preferentially consumed by gypsy moths (*Lymantria dispar*) after the seedlings were exposed to 11 episodes of 0.15 ppm O_3 for 7 hr per day over a period of 25 days (Jeffords and Endress, 1984). Oak foliage exposed to median O_3 concentrations of 0.07–0.10 ppm was less preferred than foliage exposed to lower or ambient concentrations or to artificial O_3 concentrations greater than 0.12 ppm.

Vertebrate Wildlife Consumers

Vertebrate consumers, as well as invertebrates, can affect individual processes within forest and woodland ecosystems. We know very little of the direct and indirect effects of ambient O_3 on wildlife in forest ecosystems. Reports indicate blindness in bighorn sheep, genetic changes in mice, and respiratory lesions and infections in birds, mice, rats, and rabbits (Newman and Schreiber, 1988). In the laboratory after 6-hr nighttime exposures, male rats showed a decline in physically active behaviors such as wheel running at O_3 concentrations of 0.08–0.12 ppm, and these behaviors continued to decline at higher concentrations (Tepper and Weiss, 1986). If this behavioral response also occurs in other small mammals in the field, where nocturnal O_3 concentrations remain at about 0.10 ppm (e.g., in the San Bernardino Mountains), then three ecological effects might result: 1) animals might gain less weight because of decreased seed foraging, also thereby increasing seedling germination and emergence; 2) small mammals might be more easily caught by O_3-tolerant predators, resulting in a decline in small mammal populations; or 3) both of these may occur.

However, in contrast to studies conducted on relatively pure-bred laboratory rats, some research has been done on the mortality response of wild native populations of California deer mice (*Peromyscus californicus*) captured in the Los Angeles area. They were exposed to acute O_3 doses at least 10 times higher (6.6 ppm for 12 hr) than the highest concentrations expected in the region. At these abnormally high O_3 levels, wild species appeared to be more tolerant to O_3 than mice trapped from geographic areas of low ambient air pollution.

Under field conditions in the San Bernardino Mountains of California, both the number of small mammals and the number of species caught varied across six study plots in 1972 (Kolb and White, 1974). These plots were subsequently examined intensively for vegetation injury from ambient oxidants during an ecological research program that was conducted from 1972 through 1980 (Miller et al., 1982). Kolb and White (1974) attributed the differences in small mammal populations across the study plots to vegetation differences between plots without questioning the extent to which such differences could be attributed to O_3 exposure during the preceding 20 years. Oxidant pollutant concentrations had been elevated above any realistic tropospheric background level for at least the previous 20 years in this region, and the number of small mammals and the number of species caught in 1972 were inversely related to the degree of foliar injury of yellow pine in the study plots.

Landscape-Level Ecological Responses

Another concept that evolved in North America during the 1980s was that of "landscape ecology." The focus was placed above the individual ecosystem level on spatial mosaics of habitat/ecosystem types arising from "fragmentation" across the landscape on a regional scale.

With respect to forest responses to O_3 at the level of landscape ecology, two approaches are conceivable: inductive and deductive. Even in the landscape that serves as a textbook example for O_3 effects on the forest ecosystem, the San Bernardino Mountains of southern California, higher than background levels of O_3 are known to have been present for at least four decades. Because the forest type is mixed conifer rather than a single dominant tree species, this is too short a time in the course of forest ecosystem dynamics to actually see total vegetation type conversions at the regional scale. As a result, there is no documented evidence at this time of any forest type conversions resulting from ambient O_3 exposure, and the inductive approach toward examining forest responses to O_3 at the level of landscape ecology is presently not possible.

An example of the deductive approach toward assessing forest responses to O_3 at the level of landscape ecology has been developed for Austria (Grossman and Schaller, 1986; Grossman, 1988). There are problems with this approach, especially those related to poor documentation of the model and possibly with how the relationship between O_3 exposure and forest response is quantitatively defined. However, these problems can be solved through future research.

References

Abrahamsen, G., and Stuanes, A. O. 1986. Lysimeter study of effects of acid deposition on properties and leaching of gleyed dystric brunisolic soil in Norway. Water Air Soil Pollut. 31:865-878.

Arkley, R. J. 1981. Soil moisture use by mixed conifer forest in a summer-dry climate. Soil Sci. Soc. Am. J. 45:423-427.

Arkley, R. J., and Glauser, R. 1980. Effects of oxidant air pollutants on pine litter-fall and the forest floor. Page 225 in: Proc. Symp. Eff. Air Pollut. Mediterr. Temperate For. Ecosyst. U.S. Dep. Agric. For. Serv. Gen. Tech. Rep. PSW-43.

Baltensweiler, W. 1985. "Waldsterben": Forest pests and air pollution. Z. Angew. Entomol. 99:77-85.

Bethlahmy, N. 1974. More streamflow after a bark beetle epidemic. J. Hydrol. 23:185-189.

Bethlahmy, N. 1975. A Colorado episode: Beetle epidemic, ghost forests, more streamflow. Northwest Sci. 49:95-105.

Billings, W. D. 1970. Plants, Man and the Ecosystem. Wadsworth, Belmont, CA.

Caspary, H. J. 1989. Water yield changes in catchments as an ecohydrological indicator of forest decline. Pages 111-112 in: Proc. Int. Congr. For. Decline Res.: State Knowl. Perspect. Zu beziehen bei der Literaturabteilung des Kernforschungszentrums Karlsruhe GmbH, Karlsruhe, Germany.

Coleman, J. S., and Jones, C. G. 1988. Acute ozone stress on eastern cottonwood (*Populus deltoides* Bartr.) and the pest potential of the aphid, *Chaitophorus populicola* Thomas (Homoptera: Aphididae). Environ. Entomol. 17:207-212.

Coleman, J. S., Jones, C. G., and Smith, W. H. 1987. The effect of ozone on cottonwood-leaf rust interactions: Independence of abiotic stress, genotype, and leaf ontogeny. Can. J. Bot. 65:949-953.

Committee on Forest Terminology. 1944. Forest Terminology. Society of American Foresters, Washington, DC.

Dahlsten, D. L., Rowney, D. L., and Kickert, R. N. Effects of oxidant air pollutants on western pine beetle, *Dendroctonus brevicomis* (Coleoptera: Scolytidae), populations in southern California. Can. Entomol. (In press.)

Fenn, M. E., and Dunn, P. H. 1989. Litter decomposition across an air pollution gradient in the San Bernardino Mountains. Soil Sci. Soc. Am. J. 53:1560-1567.

Friedland, A. J., Gregory, R. A., Käremiampi, L., and Johnson, A. H. 1984. Winter damage to foliage as a factor in red spruce decline. Can. J. For. Res. 14:963-965.

Führer, E. 1985. Air pollution and the incidence of forest insect problems. Z. Angew. Entomol. 99:371-377.

Grossman, W.-D. 1988. Products of photo-oxidation as a decisive factor of the new forest decline? Results and considerations. Ecol. Modell. 41:281-305.

Grossman, W.-D., and Schaller, J. 1986. Geographical maps on forest die-off, driven by dynamic models. Ecol. Modell. 31:341-353.

Guderian, R., and Kueppers, K. 1980. Response of plant communities to air pollution. Pages 187-199 in: Proc. Symp. Eff. Air Pollut. Mediterr. Temperate For. Ecosyst. U.S. Dep. Agric. For. Serv. Gen. Tech. Rep. PSW-43.

Guderian, R., and Stratmann, H. 1968. Freilandversuche zur Ermittlung von Schwefeldioxidwirkungen auf die Vegetation, vol. 3. Westdeutscher Verlag, Köln-Opladen, Germany.

Hall, R. W., Barger, J. H., and Townsend, A. M. 1988. Effects of simulated acid rain, ozone and sulfur dioxide on suitability of elms for elm leaf beetle. J. Arboric. 14(3):61-66.

Hidy, G. M., Mahoney, J. R., and Goldsmith, B. J. 1978. International Aspects of the Long-Range Transport of Air Pollutants. Final Report. U.S. Department of State, Washington, DC.

Höpker, K.-A., Führer, G., Strube, D., and Senser, M. 1989. Effects of extreme ozone concentrations on the physiology of *Picea abies* L. Karst. as related to the mineral nutrient supply of the trees. Pages 433-435 in: Air Pollution and Forest Decline.

Proc. Int. Meet. IUFRO, 14th. J. B. Bucher and I. Bucher-Wallin, eds. EAFV, Birmensdorf, Switzerland.

Jeffords, M. R., and Endress, A. G. 1984. Possible role of ozone in tree defoliation by the gypsy moth (Lepidoptera: Lymantriidae). Environ. Entomol. 13:1249-1252.

Johnson, D. W. 1986. Effects of acid deposition on forest soils. Pages 484-496 in: Proc. IUFRO World Congr., 18th. Div. 1, vol. 2. IUFRO World Organizing Committee, Ljubljana, Yugoslavia.

Johnson, D. W., Kelly, J. M., Swank, W. T., Cole, D. W., Hornbeck, J. W., Pierce, R. S., and van Lear, D. 1985. A Comparative Evaluation of the Effects of Acid Precipitation, Natural Acid Production and Harvesting on Cation Removal from Forests. Publ. ORNL-TM 9706. Oak Ridge National Laboratory, Oak Ridge, TN.

Kenk, G., Kremer, W., Bonaventura, D., and Gallus, M. 1984. Jahrring- und zuwachsanalytische untersuchungen in erkrankten tannenbeständen des Landes Baden-Württemberg. Mitt. Forstl. Versuchsanst. 12:1-37.

Kickert, R. N., and Krupa, S. V. 1990. Forest responses to tropospheric ozone and global climate change: An analysis. Environ. Pollut. 68:29-65.

Kickert, R. N., McBride, J. R., Miller, P. R., Ohmart, C. P., Arkley, R. J., Dahlsten, D. L., Cobb, F. W., Jr., Parmeter, J. R., Jr., Luck, R. F., and Taylor, O. C. 1980. Photochemical oxidant air pollution effects on a mixed conifer forest ecosystem. Final report. U.S. Environ. Prot. Agency Publ. EPA-600/3-80-002.

Kolb, J. A., and White, M. 1974. Small mammals of the San Bernardino Mountains, California. Southwest. Nat. 19:112-114.

Koth, I., and Prinz, B. 1988. Air pollution and forest decline. Research program and strategy of the state of North Rhine–Westfalia. Pages 527-536 in: Effects of Atmospheric Pollutants on the Spruce-Fir Forests of the Eastern United States and the Federal Republic of Germany. U.S. Dep. Agric. For. Serv. Northeast For. Exp. Stn. Gen. Tech. Rep. NE-120.

Krause, G. H. M., and Prinz, B. 1989. Experimental Analyses of the LIS on the Clarification of Possible Causes of the Novel Forest Decline. (In German.) LIS-Berichte 80. Landesanstalt für Immissionsschutz Nordrhein-Westfalen, Essen, Germany.

Krupa, S. V., and Kickert, R. N. 1989. The greenhouse effect: Impacts of ultraviolet-B (UV-B) radiation, carbon dioxide (CO_2), and ozone (O_3) on vegetation. Environ. Pollut. 61:263-392.

Materna, J. 1984. Impact of atmospheric pollution on natural ecosystems. Pages 397-416 in: Air Pollution and Plant Life. M. Treshow, ed. John Wiley & Sons, New York.

Materna, J. 1985. Luftverunreinigungen und waldschäden. Pages 19.1-19.3 in: Proc. Unweltschutz—Eine Int. Aufgabe. VDI, Verlag, Germany.

McLaughlin, S. B. 1985. Effects of air pollution on forests: A critical review. J. Air Pollut. Control Assoc. 35:512-534.

Miller, P. R., and McBride, J. R. 1975. Effects of air pollutants on forests. Pages 196-236 in: Responses of Plants to Air Pollution. J. B. Mudd and T. T. Kozlowski, eds. Academic Press, New York.

Miller, P. R., and van Doren, R. E. 1982. Ponderosa and Jeffrey pine foliage retention indicates ozone dose response. U.S. Dep. Agric. For. Serv. Gen. Tech. Rep. PSW-58.

Miller, P. R., Taylor, O. C., and Wilhour, R. G. 1982. Oxidant air pollution effects on a western coniferous forest ecosystem. U.S. Environ. Prot. Agency Environ. Res. Brief EPA-600/D-82-276.

Neger, F. W. 1908. Das tannensterben in den sächsischen und anderen deutschen mittelgebirgen. Tharandter Forstl. Jahrb. 58:201-225.

Newman, J. R., and Schreiber, R. K. 1988. Air pollution and wildlife toxicology: An overlooked problem. Environ. Toxicol. Chem. 7:381-390.

Olson, J. S. 1970. Geographical index of world ecosystems. Pages 297-304 in: Analysis of Temperate Forest Ecosystems. D. E. Reichle, ed. Springer-Verlag, New York.

Prinz, B. 1985. Waldschäden in den USA und in der Bundesrepublik Deutschland—Betrachtungen und Ursachen. VGB Kraftwerkstech. 65:930-938.

Prinz, B. 1987. Causes of forest damage in Europe. Environment 29(9):11-37.

Roberts, T.-M., Skeffington, R. A., and Blank, L. W. 1989. Causes of type I spruce decline in Europe. Forestry 62(3):179-223.

Runeckles, V. C. 1986. Photochemical oxidants. Pages 265-303 in: Air Pollutants and Their Effect on the Terrestrial Ecosystem. A. H. Legge and S. V. Krupa, eds. John Wiley & Sons, New York.

Schäfer, H., Bossel, H., Krieger, H., and Trost, N. 1988. Modeling the responses of mature forest trees to air pollution. GeoJournal 17:279-287.

Schulze, E.-D. 1989. Air pollution and forest decline in a spruce (*Picea abies*) forest. Science 244:776-783.

Schütt, P., and Cowling, E. B. 1985. Waldsterben, a general decline of forests in central Europe: Symptoms, development, and possible causes. Plant Dis. 69:548-558.

Seufert, G., Arndt, U., Jäger, H.-J., and Bender, J. 1989. Long-term effects of air pollutants on forest trees in open-top chambers. I. Experimental approach and results on mineral cycling. Pages 159-165 in: Air Pollution and Forest Decline. Proc. Int. Meet. IUFRO, 14th. J. B. Bucher and I. Bucher-Wallin, eds. EAFV, Birmensdorf, Switzerland.

Solar, M. 1986. Air pollution and forest decline in Slovenia. Pages 368-380 in: Proc. IUFRO World Congr., 18th. Div. 2, vol. 1. IUFRO World Congress Organizing Committee, Ljubljana, Yugoslavia.

Svensson, G. 1984-1985. Forest damage surveyed. Acid Mag., Winter, pp. 12-13.

Tepper, J. S., and Weiss, B. 1986. Determinants of behavioral response with ozone exposure. J. Appl. Physiol. 60:868-875.

Treshow, M. 1980. Interactions of air pollutants and plant disease. Pages 103-109 in: Proc. Symp. Eff. Air Pollut. Mediterr. Temperate For. Ecosyst. U.S. Dep. Agric. For. Serv. Gen. Tech. Rep. PSW-43.

Ulrich, B. 1984. Langzeitwirkung von luftverunreinigungen auf waldökosysteme. Düsseldorfer Geobot. Kolloq. 1:11-24.

Wetstone, G. 1983. A conversation with acid rain researcher Hubert Vogelmann. Environ. Forum, February, pp. 21-28.

Wieler, A. 1911. Pflanzenwachstum und Kalkmangel im Boden. Bornträger, Berlin.

Wislicenus, H., ed. 1907. Sammlung von Abhandlungen über Abgase und Rauchschäden. Paul Parey, Berlin.

Zinke, P. J. 1980. Influence of chronic air pollution on mineral cycling in forests. Pages 88-99 in: Proc. Symp. Eff. Air Pollut. Mediterr. Temperate For. Ecosyst. U.S. Dep. Agric. For. Serv. Gen. Tech. Rep. PSW-43.

Further Reading

Chevone, B. I., and Linzon, S. N. 1988. Tree decline in North America. Environ. Pollut. 50:87-99.

McLaughlin, S. B. 1985. Effects of air pollution on forests: A critical review. J. Air Pollut. Control Assoc. 35:512-534.

Miller, P. R. 1993. Response of forests to ozone in a changing atmospheric environment. Angew. Bot. 67:42-46.

Prinz, B., Krause, G. H. M., and Jung, K.-D. 1987. Development and causes of novel forest decline in Germany. Pages 1-24 in: Effects of Atmospheric Pollutants on Forests, Wetlands and Agricultural Ecosystems. T. C. Hutchinson and K. M. Meema, eds. Springer-Verlag, New York.

Skelly, J. M., and Innes, J. L. 1994. Waldsterben in the forests of central Europe and eastern North America: Fantasy or reality? Plant Dis. 78:1021-1032.

Smith, W. H. 1990. Air Pollution and Forests: Interactions Between Air Contaminants and Forest Ecosystems. Springer-Verlag, New York.

CHAPTER 9

Control Strategies for Air Pollution

> In 1843, there was another Parliamentary Select Committee, and in 1845, a third. In that same year, during the height of the great railway boom, an act of Parliament disposed of trouble from locomotives once and for all . . . by laying down the dictum that they must consume their own smoke. The Town Improvement Clauses Act two years later applied the same panacea to factory furnaces. Then 1853 and 1856 witnessed two acts of Parliament dealing specifically with London and empowering the police to enforce provisions against smoke from furnaces, public baths, and washhouses and furnaces used in the working of steam vessels on the Thames.
>
> —Sir Hugh Beaver

Introduction

Societal concern for air quality has been known since the times of Seneca, 61 A.D. (see Chapter 3, An Historical Perspective of Air Quality). In 1845, public laws were enacted in England to control air pollution from both mobile and stationary sources (Beaver, 1955). The industrial revolution of the 20th century simply contributed to the disturbing atmospheric burden of human-made pollutants through massive growth of industries with diverse arrays of pollutant emissions and added to the technological complexities associated with their control and the corresponding economic and social costs.

Since the 1970s, there has been a significant growth in public awareness of measures to improve air quality and consequently to protect human health and the environment. These considerations have been the domain predominantly of the developed nations, while developing countries have struggled with issues such as population growth, illiteracy, and starvation. In the so-called new world order, environmental quality and protection at the global scale will involve not only economic cooperation between developed and developing nations, but also technology transfer and a sharing of knowledge and experience between these nations. Godish (1991) has provided a thorough summary of control technologies for air pollutants.

Air pollution control and the resulting improvement in air quality have been achieved through the application of industrial engineering technology and the implementation of regulatory policies (emission caps or limits on pollutants and air quality standards). Both approaches have been difficult to implement in developing nations (e.g., see the narrative on air pollution in Mexico City in Chapter 3). Nevertheless, emissions from mobile and stationary sources require different types of control strategies (Tables 9.1–9.3). A technological consideration here is that the engineering design of the emission source must be compatible with the engineering design of the control technology. For example, cars built prior to 1974 such as the Volkswagen Beetle cannot be equipped with modern emission control technology such as the catalytic converter because of their engine design. This is also true for stationary sources.

Table 9.1. Air pollution control strategies for mobile sources (motor vehicles)

Problem	Solution
1. Crankcase emissions	Recirculation
2. Carburetor and fuel tank evaporative emissions	Charcoal filter
3. Exhaust gases	
a. Carbon monoxide, hydrocarbons	Catalytic converter
	Oxygenated fuels
b. Oxides of nitrogen	Spark ignition timing
	Air–fuel mixture compression ratio
	Exhaust gas recirculation

Table 9.2. Air pollution control strategies for stationary sources

1. Modification of plant operating procedures
2. Use of tall stacks
3. Emission control
   - A. Particulate matter
     - a. Cyclone
     - b. Fabric filtration
     - c. Electrostatic precipitation
     - d. Wet scrubbing
       - i. Open spray tower
       - ii. Venturi scrubber
   - B. Gases
     - a. Combustion
       - i. Direct flame incineration
       - ii. Thermal incineration
       - iii. Catalytic incineration
     - b. Adsorption
     - c. Absorption

Table 9.3. Some examples of air pollution control strategies for stationary sources and their applications

Control strategy	Applications
Adsorption (solvent recovery)	Dry cleaning; surface coating; rayon, plastic, and rubber processing; control of offensive odors
Catalytic incineration	Varnish cooking; paint and enamel baking; printing presses; coke ovens
Direct flame incineration	Petrochemical plants and refineries
Electrostatic precipitation	Sulfuric acid plants; metallurgical operations; steel-making processes; coal-fired power plants
Fabric filtration	Cement kilns; foundries; oil-fired boilers; carbon black plants; steel-making operations; coal-fired power plants that use low-sulfur coal
Tall stacks	Coal-fired power plants; metal smelters
Thermal incineration	Varnish cooking; paint baking; meat smokehouses; resin manufacturing
Wet scrubbing	Coal-fired power plants

New sources in the United States must comply with the New Source Performance Standards (NSPS), while such regulatory policies are not fully implemented in many developing nations, primarily because of the associated high economic costs.

Air Pollution Control Through Process or Engineering Technology

Air Pollution Control Strategies for Mobile Sources (Motor Vehicles)

Crankcase Emissions

Crankcase emissions can be controlled by a relatively simple and inexpensive technology called positive crankcase ventilation, in which emission gases are recirculated to the combustion chamber for reburning.

Carburetor and Fuel Tank Emissions

Emissions from the carburetor and fuel tank can be controlled by the use of a charcoal canister mounted under the vehicle's hood. When the engine is not operating, gasoline evaporating from the carburetor and the fuel tank is collected and stored in the charcoal canister. On engine start, warm air is drawn through the charcoal, purging it of collected gasoline vapors, which are then drawn into the combustion chamber.

Carbon Monoxide (CO) and Hydrocarbons (HCs)

During the early days of emission control, CO and HCs were reduced by a variety of engine modifications designed to improve combustion. In contrast, catalytic converters are after-devices, reducing emissions by combusting exhaust gases after they leave the combustion chamber. The converter is mounted in front of the muffler and close to the engine exhaust manifold.

The catalyst helps O_2 to combine with CO and unburned HCs to produce CO_2 and H_2O. Although the catalytic materials participate in combustion reactions, they do not themselves undergo chemical conversion. In catalyzing the combustion, the catalytic converter acts as an incinerator. However, this incineration occurs at much lower temperatures (350–400°C) than those required for thermal incineration (700–750°C) and consequently does not require a supplemental energy source for efficient combustion. As combustion takes place in the converter, temperatures may rise to more than 500°C.

Two basic geometric configurations have been used in catalyst design: the monolith and the pellet. Both designs use the noble metals platinum and palladium as catalysts, either individually or in a platinum-palladium blend. Although these are more expensive than other catalyst materials, they have the advantage of not being "poisoned" by the sulfur compounds in gasoline. Because of their catalytic efficiency, less of these metals is required per vehicle.

The monolithic design consists of a cylinder 7.6–15.2 cm in diameter, which has an internal ceramic honeycomb structure coated with a highly catalytic material. In the pellet design, small pellets are impregnated with catalyst materials. In both designs, there are thousands of passages that allow exhaust gases to flow freely through the converter with relatively low back pressure. The high surface area, >1,000 m^2, allows the catalyst materials to come into direct contact with the exhaust gases, oxidizing them to CO_2 and water vapor.

As stated previously, in order for catalysts to function efficiently, it is essential that they not be contaminated by chemicals that may poison them and render them inactive. For this reason, motor vehicles with catalytic converters cannot use fuels that contain lead and phosphorus. Anticipating that auto manufacturers would employ catalyst technology for 1975-model vehicles, the U.S. EPA required petroleum companies to make unleaded, phosphorus-free fuels available for distribution. To prevent consumer use of leaded fuels in catalytic systems, the EPA also required auto manufacturers to employ small-diameter fuel tank filler necks and fuel distributors to use small-diameter gasoline pump nozzles.

If leaded gasoline were used in a catalyst system, either accidentally or deliberately, the system would be rendered inactive, since lead will coat the catalyst and make it inaccessible to exhaust gases. A few tanks of leaded gasoline should not destroy the catalytic converter; it would, however, reduce the system's effectiveness for a period of time. When unleaded gasoline is again used exclusively, catalyst performance should return to normal.

In addition to the converter itself, such systems also utilize a quick heat intake manifold, electronic ignition, and a computer-controlled fuel supply system. However, catalytic converters are least effective during cold-start and engine warm-up. A rich air-fuel mixture must be used during frigid temperatures to compensate for the low volatility of cold fuel. This rich fuel mixture is not completely combusted, and high CO and HC emissions result. To reduce cold-start emissions, auto manufacturers have employed leaner air-fuel mixtures, preheating of air and fuel, improvement of air-fuel mixture control through the use of fuel injectors, and shorter choking periods.

In the quick heat intake system, air is drawn over the exhaust manifold where it is warmed to simulate summer ambient temperatures. The heating of the intake air improves fuel vaporization. The higher volatility of the fuel allows for leaner air-fuel engine operation and a shorter choking period during warm-up. Recently in the United States, vehicles have been required to use oxygenated or oxygen-rich fuels (e.g., gasoline blended with ethanol). Use of such fuels is thought to result in more complete combustion, producing CO_2 and H_2O rather than the normally high concentrations of CO and HCs produced by the hard stops and starts of vehicles, particularly during very cold periods. This approach is primarily directed at reducing CO emissions, since the emission of HCs during the winter does not lead to high O_3 production because of the low solar radiation (i.e., reduced photochemical processes).

The ideal air-fuel ratio (called the stoichiometric ratio) for unleaded gasoline is 14.7:1. At this ratio, all available fuel and oxygen will be consumed, efficiency and power will be greatest, and emissions of pollutants such as oxides of nitrogen (NO_x) and unburned HCs will be lowest. The addition of ethanol to gasoline is expected to reduce HC emissions from automobile engines. However, since ethanol is a less efficient fuel than gasoline, fuel mileage is decreased. In modern fuel-injected engines, this leaner ethanol-gasoline mixture causes no additional problems because an electronic engine-management system adjusts the mixture dynamically while the engine is operating. In engines with carburetors, the mixture is set by the size of jets, valves, the venturi, and hard parts in the carburetor itself. An engine that operates well on pure gasoline will therefore run "lean" on ethanol-gasoline mixtures, resulting in poor idle (releasing unburned HCs), poor mileage and performance, and a hotter combustion temperature (which increases NO_x emissions). In addition, ethanol may damage certain parts of fuel systems. Many vehicle manufacturers specifically prohibit the use of ethanol-containing fuels in certain lines of their engines and exclude this type of damage from new car warranty repair. The issue is politically charged, because ethanol is produced from corn and seen as a boon to farmers. In reality, the current market structure for corn in the United States makes the value of this strategy questionable at best.

Oxides of Nitrogen (NO_x)

Control of nitrogen oxides emission has lagged behind that of CO and HCs. The control of NO_x poses a dilemma since control measures employed to decrease emissions of CO and HCs increase the efficiency of combustion. As combustion efficiency increases, higher combustion chamber temperatures result, which enhance the formation of NO_x. As previously discussed, combustion efficiency can be improved by using lean air-fuel mixtures. Nitrogen oxides, on the other hand, can be significantly reduced by utilizing rich air-fuel mixtures. These rich mixtures result in lower combustion chamber temperatures and decreased emissions of NO_x, but CO and HC emissions are increased significantly.

Historically, control measures for NO_x have employed techniques that reduce combustion chamber temperatures without significantly increasing emissions of CO and HCs. These have included the retardation of spark timing, decreased compression ratios, and exhaust gas recirculation.

Spark Ignition Timing

At highway speeds, spark ignition normally takes place at 15–20 degrees before the piston reaches top dead center. Retardation of the spark so that ignition occurs later in the compression stroke reduces the time available for the production of higher temperatures, resulting in decreased NO_x production. As the spark is retarded, the surface area-to-volume ratio of the combustion chamber is also decreased. Since there is less wall surface available for quenching the flame front, HC emissions are also decreased.

Air-Fuel Compression Ratio

Before ignition, the air-fuel mixture is compressed by the upward motion of the piston. At maximum compression, the air volume of the combustion chamber is only a fraction of that which existed before compression was initiated. Therefore, the compression ratio is the relationship between the initial and the final combustion chamber volumes.

Lowering compression ratios produces lower combustion chamber temperatures. Vehicles produced before the early 1970s had relatively high compression ratios, about 8.5:1 to 11.1:1. Modern-day compression ratios have been lowered to about 8.1:1 to 8.5:1 to reduce NO_x emissions and to facilitate the use of lower octane, unleaded fuels required for the effective operation of catalyst-equipped vehicles.

Exhaust Gas Recirculation

One of the most effective ways of reducing NO_x emissions has been exhaust gas recirculation (EGR), in which controlled amounts of exhaust gas are recirculated back to the intake manifold, where it mixes with air and fuel entering the combustion cycle. These recirculating gases are relatively inert (i.e., they contain very little combustible material). Carbon dioxide and water vapor in this recirculated exhaust gas serve as heat sinks, reducing peak combustion temperatures.

Air Pollution Control Strategies for Stationary Sources

Modification of Plant Operating Procedures

Although it is often assumed that air pollution control requires the use of gaseous pollutant-controlling devices, much progress in achieving air quality standards has resulted from the implementation of a variety of changes in plant operating procedures. These include the pretreatment of process materials, material substitution, and process changes.

Significant reductions in emissions of particulate matter and sulfur from coal boilers can be achieved by coal washing, which reduces not only the ash content of the coal, but also inorganic sulfur. A significant desulfurization of oil can also be achieved with the addition of several steps to petroleum-refining processes.

Emission reduction may also be achieved by substituting materials that perform equally well in a process. The use of cleaner-burning fuels (natural gas and low-sulfur fuel oil or coal in place of high-sulfur coal) has been a major factor in the significant reduction of ambient SO_2 and particulate matter levels in many of the air quality control regions in the United States. These fuel substitutions have not, however, been problem free. Low-sulfur fuels are in short supply, and distributors must import low-sulfur oil, increasing the reliance on foreign suppliers. The use of low-sulfur western coal has reduced the markets for high-sulfur eastern coal in the United States. Fuel substitutions have paradoxically contributed to improving the quality of urban air while they have exacerbated the energy-supply problems in the United States.

Significant improvements in air quality can also be achieved by producing electricity from nuclear materials rather than from fossil fuels. Indeed, the marked improvement in air quality in Chicago has resulted not only from the use of low-sulfur fuels, but also from the use of nuclear power to produce ~40% of the city's electric power. Of course, as the accidents at Three Mile Island in Pennsylvania (1979) and Chernobyl in the former USSR (1986) illustrated, nuclear power has its own unique environmental hazards.

Source emissions may also be reduced by changes in processes and operating procedures. For example, the loss of volatile materials in chemical and petroleum industries can be minimized by processing materials in completely enclosed systems. In addition, vapors from volatile petroleum product storage tanks can be condensed and reused.

Use of Tall Stacks

Historically, smokestacks have been used to reduce pollution problems by elevating emissions above the ground where they may be more effectively dispersed. In the absence of control devices and because of public concern for air quality, the use of smokestacks was a relatively effective means of reducing ground-level concentrations. The effectiveness of a smokestack in pollutant dispersion depends in part on stack height, plume velocity and temperature, atmospheric conditions such as wind speed and direction, atmospheric stability, topography, and location of other sources.

The use of tall stacks (200–400 m) to reduce ground-level concentrations of SO_2 from utility boilers has been the preferred method of emission control in coal-fired electric power plants. The higher wind speed and greater air volume available at increased elevations result in significant dilution of pollutants. Although tall stacks effectively reduce ground-level concentrations, they do so by dispersing pollutants over a wider area. Therefore, the use of tall stacks does not result in a net reduction in emissions. Furthermore, the use of tall smoke stacks has been criticized for contributing to the widespread occurrence of acidic rain.

Particle Collection Systems

Cyclones

Particles can be removed from a waste gas stream by the induction of cyclonic airflow. In a common cyclone, waste gas enters a tangential inlet near the top of a cylindrical body (Fig. 9.1A). This creates a main vortex that spirals downward between the walls of the cyclone and the centrally placed discharged outlet. Airflow continues downward below the walls of the gas outlet until it reaches the bottom of a cone, where the vortex changes its direction of flow, forming an inner vortex traveling upward to the gas outlet. The centrifugal forces induced by

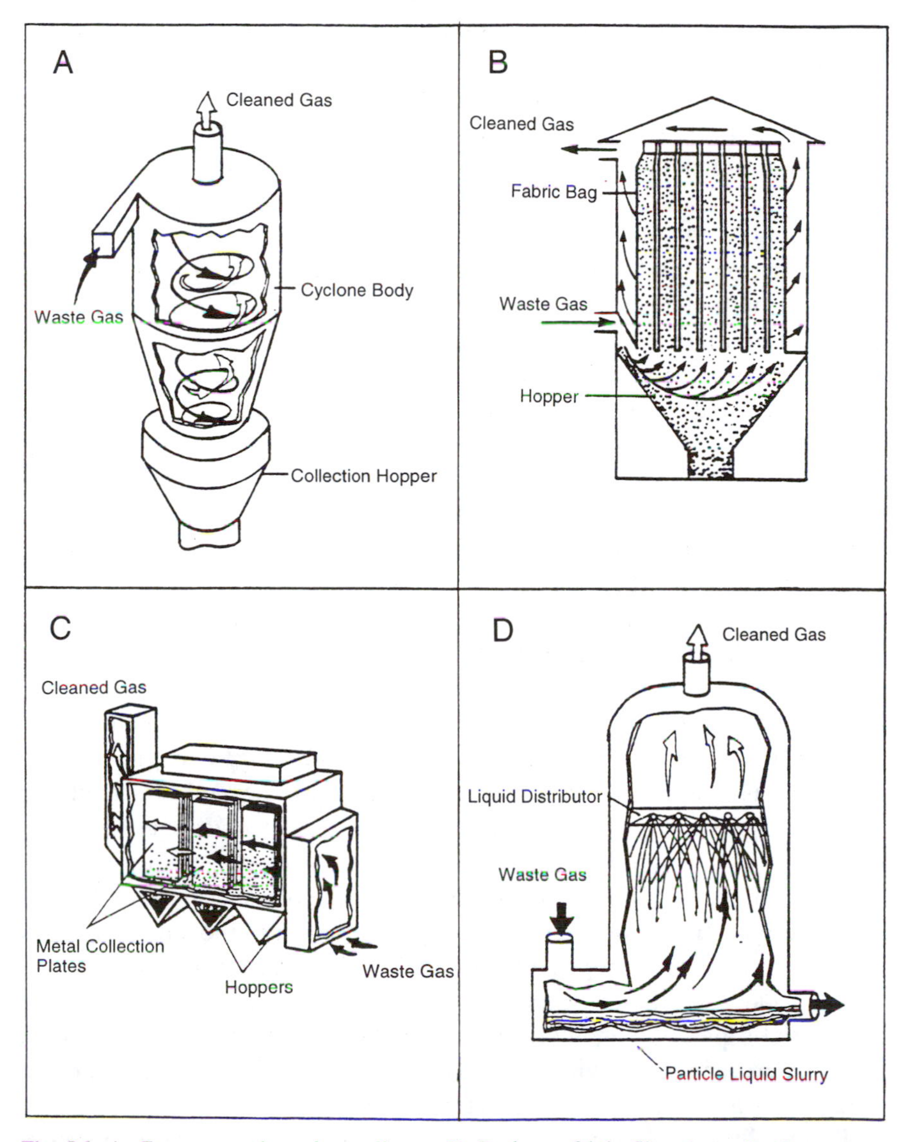

Fig. 9.1. A, Common cyclone dust collector. **B,** Baghouse fabric filter dust-collection system. **C,** Electrostatic precipitator. **D,** Open spray tower scrubber system. (Adapted from Godish, 1991)

the main vortex cause the inertial impaction of particles on the walls of the cyclone. The concentrated dust layer formed on the walls swirls downward to a hopper, where the dust is removed.

Cyclones may be classified as conventional or high efficiency. The latter have body diameters up to 25 cm, which allow them to achieve greater separating forces. The efficiency of cyclonic dust collectors may be determined by particle size or density, the velocity of the incoming gas stream, dust loadings, and equipment design parameters such as the ratio of the body diameter to gas outlet diameter and the cyclone body or cone length. In general, particle collection efficiency increases with increased particle size and density, dust loading, and increased collector size and design parameters.

Filtration

In filtration, solid particles are physically collected on fibers on the inside of bags (fabric filters) or on filter beds (Fig. 9.1B). Filtration is one of the most reliable and efficient methods for dust collection, exceeding 99.5% collection efficiencies in some systems.

Fabric filters. Fabric filters are employed to control dust emissions from a variety of industrial sources including cement kilns, foundries, oil-fired boilers, carbon black plants, and electric and oxygen furnaces for steel-making operations. They are also used to collect fly ash from electricity generating plants that burn low-sulfur coal. Fabric filters are commonly used for the control of particle emissions where dust loadings are high, particle sizes are small, and high collection efficiencies are required. Fabric filters are designed to handle gas volumes of 280–1,400 m^3 per minute.

The fabric filter collection system consists of multiple tubular (Fig. 9.1B) or flat collecting bags suspended inside a housing. A single housing, called a baghouse, may contain several hundred to several thousand bag filters. Bags are made from a variety of fabrics. Choice of the fabric depends on temperature, moisture, and chemical composition of the gas, as well as on the physical and chemical nature of the particles to be collected.

In the typical baghouse unit, waste gas enters from the side and flows downward toward a hopper, where the flow is reversed upward into the array of bags. As the gas changes direction, large particles are removed by inertial separation and collected in the hopper. As gas passes through the tubular bags, dust collects on the inside of the bag surface, and the filtered gas is discharged into the atmosphere.

Fabric filtration is similar to the process employed in a home vacuum cleaner, except that positive, rather than negative, air pressure causes dirty air to pass into the collection bag. Like a home vacuum cleaner, collected dust must also be periodically removed. Bags may be cleaned by mechanical shaking, which is the most commonly used cleaning method.

Dust collection on bag filters involves more than the collection of particles on single fibers, since open spaces between fibers are often much larger than the particles to be collected. Efficient collection of small particles is accomplished by the formation of a particle mat across the filter fibers. Collection efficiencies are lowest when filters are first installed and immediately after cleaning. Maximum efficiency occurs when mat buildup has taken place.

Although fabric filters provide high overall collection efficiencies (>99%) and represent the most promising technology for controlling fine particles, they do have limitations. These include high capital costs, flammability hazards for some types of dusts, large space requirements, flue gas temperatures limited to 285°C, and sensitivity to moisture content in the gas stream. Because of the abrasiveness of dust and the bags' sensitivity to chemicals, bag wear is a major maintenance consideration. Average bag life is about 18 months, although many industries change them once a year.

Electrostatic Precipitation

Electrostatic precipitators remove solid and/or liquid particles from effluent gases by imparting a charge to entrained particles, which are then attracted to positively charged collection plates (Fig. 9.1C). Specifically, high DC voltage is applied to a dual-electrode system that produces a corona between the discharge electrode and the collection plates or electrodes. The corona produces negatively charged ions, and particles entrained in the gas stream become negatively charged as a result. These negatively charged particles then migrate toward the positively charged plates, where they are collected and held by electrical, mechanical, and molecular forces. Solid particles are usually removed by periodic rapping of collection electrodes; collected dust falls into a hopper and then is manually removed.

The electrostatic precipitator was developed in 1907 to collect acid mist from sulfuric acid plants. Historical applications have been the collection of metal oxides and dusts from a variety of metallurgical operations, including ferrous and nonferrous processes. It is commonly used in steel making to collect particles from blast, basic O_2, and open hearth furnaces, sintering plants, and coke ovens. Presently, the largest single application of electrostatic precipitation is the collection of fly ash from boilers in coal-fired power plants.

Collection efficiency is primarily influenced by particle retention time in the electrical field and particle resistivity. Retention time is determined by gas path length (the distance gas travels across the electrical field) and gas velocity. The highest collection efficiencies are achieved in precipitators employing long gas path lengths and low gas velocity.

The resistivity of particles to an electrical charge is one of the most serious problems affecting collection efficiency. When dust resistivity is high, as in fly ash from low-sulfur western coals, the charge is not neutralized at the collection plate; as a consequence, an electrical potential builds up on the collected dust. As the potential increases, the incoming particles do not receive a maximum charge, and the collection efficiency is decreased. If this condition continues, a corona discharge may occur at the collection plate, causing the collector to malfunction. Paradoxically, low-sulfur coals, which are widely used in coal-fired utility boilers to comply with SO_2 emission standards, produce fly ash of high resistivity. On the other hand, high-sulfur coals, which were used prior to the regulation of SO_2 emissions, have low fly ash resistivity and therefore high collection efficiency.

Thus, resistivity is a direct function of the sulfur content of the coal, with low-sulfur coals producing fly ash with the highest resistivity. Control methods employed for high-resistivity dusts require a high level of technical sophistication, and of course the cost is also high. As a result of the resistivity problem, fabric filtration is used as an alternative to the electrostatic precipitator for utility boilers using low-sulfur coal.

At the other end of the spectrum are dusts with low resistivity, such as carbon blacks. In such cases, collection efficiency is also markedly decreased because carbon particles are highly conductive and lose their charge too readily. Because the collected particle loses its negative charge, it may reenter the waste stream and begin the charging process again. It becomes negatively charged and is again attracted to the collection plate. The process is repeated until the particles reach the precipitator outlet, where they are discharged into the atmosphere. Because of this phenomenon, low resistivity also results in an excessive dust reentrainment.

Electrostatic precipitation is a widely used dust-collection method because it has a high collection efficiency for all particle sizes, including fine particles. Collection efficiencies in excess of 99% can be achieved, even on very large gas flows. Electrostatic precipitators have low operating and power requirements because energy is necessary to act only on particles collected and not on the entire waste stream. Consequently, energy costs are low compared with other dust-collection systems.

Although electrostatic precipitation has advantages that make it well suited to many dust-collection applications, it also has disadvantages. In addition to the resistivity problem, these include high capital costs and large space requirements.

Wet Scrubbing

Dust collection by wet scrubbing systems requires the introduction of a liquid into the effluent stream and the subsequent transfer of particles to the scrubbing liquid. This transfer is accomplished primarily by conditioning, which increases particle size, and secondarily by the entrapment of particles on a liquid film. Both of these transfer mechanisms are used in commercially available wet scrubber systems.

Scrubber designs vary considerably from one manufacturer to another. However, all scrubbers have two basic components: a section where liquid and gas contact occurs and a deentrainment section where wetted particles are removed. In wet scrubbers, particles are brought into contact with the liquid to form a particle-liquid agglomerate. The contact process is achieved by forcing a collision between the liquid and the particles. Collisions may be promoted by gravity, impingement, and mechanical impaction. When contact is made, the mass and size of the particles significantly increase. The resultant particle-gas agglomerations are removed by inertial devices. Commonly used deentrainment mechanisms include impaction on extended baffles and centrifugal separation. For simple dust-collection systems, the contact liquid is usually water.

Open spray tower. In the open spray tower, a scrubbing liquid is sprayed downward through low-pressure nozzles (Fig. 9.1D). The dust-laden waste gas enters from the side of the tower and moves downward toward the liquid pool at the bottom of the tower. Large particles are removed by impingement on the liquid surface. The waste air then changes direction and moves counter to the flow of the scrubbing liquid and is discharged at the top. Dust particles are captured by the falling droplets, and the liquid-particle agglomerate is collected in the liquid pool at the bottom of the tower. Use of spray towers is limited to low gas flow systems to avoid liquid drop entrainment in the scrubbed gas stream. Such scrubbers attain relatively low efficiencies (80–90%) and are usually employed as a precleaning system to remove particles larger than 5 μm in diameter.

Venturi scrubber. The Venturi scrubber is used when high collection efficiency is desired. Effluent gases enter the venturi section, where they impinge on the wetted cone and throat (Fig. 9.2A). As waste gas enters the annular orifice of the venturi, its velocity increases. This high velocity results in a shearing action that atomizes the scrubbing liquid. The high differential velocity between the gas stream and the liquid promotes the impaction of dust particles on the droplets. As the gas leaves the venturi

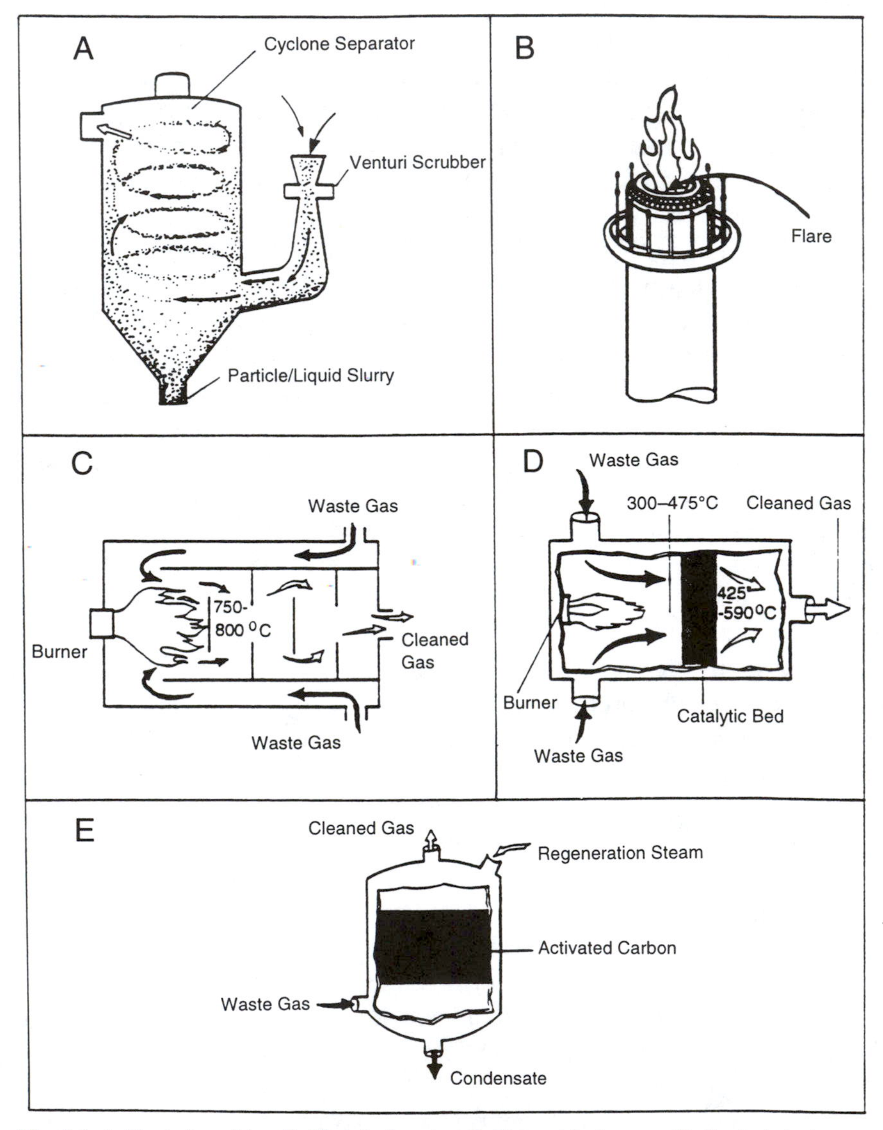

Fig. 9.2. A, Venturi scrubber. **B,** Flare incinerator. **C,** Thermal incinerator. **D,** Catalytic incinerator. **E,** A type of adsorption system. (Adapted from Godish, 1991)

section, it decelerates and impaction and agglomeration of particles and liquid continues. Particle-liquid droplets are removed from the waste system by centrifugal forces in a cyclonic deentrainment section.

Wet scrubbers can provide high collection efficiencies in a variety of applications. Because they are effective in controlling fine particles, they are almost exclusively used for such applications. Venturi scrubbers, for example, are used to collect particles from basic oxygen steel-making processes.

Although capital costs are relatively low, operating and maintenance costs are often very high. These high operating costs are primarily the result of energy requirements needed to provide high collection efficiency for fine particles. Additional operating costs are associated with the disposal of collected dusts and waste liquid. High maintenance requirements often result from the erosion and corrosion of scrubber surfaces by the abrasion and action of corrosive acids.

Gas Control Systems

Combustion

Direct flame incineration. Direct flame systems are employed when effluent gases are between the **LEL** (lower explosion limit) and **UEL** (upper explosion limit). Direct flame systems such as the flare incinerator are used in petrochemical plants and refineries producing combustible waste gases that cannot be conveniently treated any other way. The flare unit is an open-ended combustion system pointed skyward and is readily visible to the public (Fig. 9.2B).

Thermal incineration. Thermal incineration is normally utilized for the treatment of effluent gases that have a combustible concentration below the LEL. Thermal incineration requires that effluent gases be heated to high temperatures (750–800°C). Efficient combustion can be achieved by providing the necessary reaction time, optimal retention or residence time, and turbulent mixing of reaction gases (Fig. 9.2C).

Efficient thermal incineration requires supplemental fuel to bring reaction chamber temperature to 750–800°C. The amount of supplemental fuel needed can be reduced considerably by passing hot exhaust gas through a heat exchanger, which extracts heat energy from the exhaust gas and transfers it to the gases entering the combustion chamber. The heat exchanger may provide up to 75% of the energy necessary for the incineration of waste gases.

Thermal incineration is used to treat effluent streams from a number of industrial processes including varnish cooking, paint bake ovens, meat smokehouses, fat-rendering plants, paint application, and resin manufacturing. The major advantage of thermal incineration in these applications is its dependable performance. However, because of fuel requirements, operating costs are high.

Catalytic incineration. Effluent gases with combustible concentrations below the LEL can also be treated by incineration that employs catalytic substances to combust waste gases. The effluent gas stream is initially heated to temperatures of 300–475°C in a gas-fired, preheated chamber (Fig. 9.2D). The preheated gases are discharged to a catalytic bed, where combustion occurs. The most effective and commonly used catalysts are platinum and palladium, although other metals or metal salts may be used. As effluent gases enter the catalytic bed, they are adsorbed on the surface of the catalyst, increasing the concentration of reacting substances. Combustion efficiencies of 95–98% can reasonably be expected in industrial applications. The heat released may increase the temperature of treated gases by 100–300°C. As in thermal incineration, these hot exhaust gases can be passed through a heat exchanger to provide some of the preheat exchange necessary to promote efficient combustion.

Catalytic incineration has several advantages compared with thermal incineration. First, retention times for effluent gases in catalytic units are significantly lower. The retention requirement for thermal incineration is about 20–50 times greater. As a consequence, catalytic incinerators are much smaller than thermal units, and the smaller size means lower capital equipment costs. The most attractive advantages, however, are the lower supplemental fuel requirement and operating costs. Despite these advantages, catalytic incineration does not have the wide acceptance and use of thermal incinerators, primarily because catalytic systems may be poisoned by a variety of substances, including lead, phosphorus, zinc, and iron oxide. This poisoning or fouling of catalytic surfaces decreases performance and increases catalyst maintenance.

Catalytic incineration is used to treat effluent gases from a variety of industrial processes, including emissions from varnish cooking, paint and enamel bake ovens, asphalt oxidation, phthalic anhydride manufacturing, printing presses, coke ovens, and the making of formaldehyde.

Adsorption

Some contaminant gases can be removed from effluents by their physical adsorption to solid surfaces (Fig. 9.2E). The solid collecting media are called **adsorbents** and the collected gases or vapors the **adsorbate**. The adherence of adsorbate to adsorbents results from the van der Waal's intermolecular attraction forces between them. Adsorption is reversible; the adsorbate can be removed from the adsorbent by increasing the temperature or lowering the pressure. Because of this reversibility, adsorption is widely used for solvent recovery in dry cleaning, metal degreasing operations, surface coating, and rayon, plastic, and rubber processing. The application of adsorption technology has received only limited use in solving ambient air pollution problems. The primary air pollution application has been in the control of offensive odors. Other applications include the collection of noncombustibles (contaminants that are difficult to burn) and gases that cannot be easily collected by other methods.

Efficient adsorption of gases and vapors can be achieved by adsorbents that have a large surface area and that preferentially attract a specific adsorbate. Activated carbons have a very large surface-to-volume ratio and are commonly used as industrial adsorbents for the collection of organic compounds. They can be produced from various materials, including coal, coke, coconut char, charcoal, lignite, and bone char. These activated carbons vary in their affinity for different gases and vapors and in their adsorption capacities. For organic compounds, adsorption capacity increases with molecular weight. The larger, heavier substances are more completely adsorbed.

Although adsorption efficiencies of 95–99% can be easily achieved, the actual adsorption efficiency will depend on several operating parameters including gas flow rate, temperature, and adsorbent saturation. Efficient treatment requires sufficient retention time in order for the adsorbate and adsorbent to make optimal contact. Low gas velocity can increase the retention time, but it must also be low enough to prevent the movement of the adsorbent in a fixed bed. As the adsorbent approaches saturation with adsorbate, collection efficiency decreases. To maintain high efficiency, the saturated adsorbent unit must be regenerated.

Absorption

Control systems that employ liquid media to remove gases are called scrubbers. Scrubbers remove gases by chemical absorption in a medium that may be a liquid or a liquid-solid slurry. Because of its low cost, water is the most commonly used scrubbing medium. Additives, however, are commonly employed to increase chemical reactivity and absorption capacity. For example, lime or ethanolamine is added to scrubbing media for the control of acidic gases.

Scrubbers must be designed to provide maximum contact between the gas phase and scrubbing medium. Significant gas-media contact can be achieved by various mixing mechanisms, including impingement, spraying, atomization, and agitation. Gas-media contact can also be enhanced by the use of column or tower packing materials, which are usually ceramic and made in a variety of configurations designed to provide a large amount of surface area.

Packed towers. The most widely used scrubbing system is the fixed-bed packed tower. The scrubbing medium is introduced at the top and flows downward, counter to the flow of the effluent gas. The efficiency of absorption is affected by the height of the packed tower, since the longer the gas path through the scrubbing medium, the greater the probability that contaminant gases will be transferred and absorbed. The cross-sectional area of the tower determines capacity of gas flow rate; design flow rates range from 28 to 280 m^3 per minute.

Sulfur Oxides Control Systems

A variety of SO_x control technologies have been developed or are presently in the developmental stages. Coal beneficiation, solvent refining, and coal gasification remove sulfur prior to combustion. Others, such as fluidized-bed combustion, remove sulfur during the fuel-combustion process, and flue gas desulfurization systems remove SO_x after combustion but before the gases are emitted into the atmosphere.

Coal Beneficiation

Coal is a heterogeneous substance and is a composite of combustible organic and inorganic mineral matter. Both coal fractions contain sulfur. Pyritic sulfur, as well as other mineral matter, is removed by differences in specific gravity. Cleaning in which coal particles are suspended in water removes an average of 50% of the pyritic sulfur and 20–30% of the total sulfur.

Although coal beneficiation is commonly employed by coal producers to reduce ash-forming substances and has considerable potential to reduce sulfur levels in coal, it usually has not been employed to comply with SO_x emission limitations. This is partly because sulfur reduction by beneficiation is insufficient to meet emission standards. Coal beneficiation may, however, prove useful when combined with other sulfur-control technology, such as flue gas desulfurization.

Solvent Refining

Solvent refining, or chemical cleaning, removes sulfur and other impurities by dissolving coal in organic solvents. This technology is still in the pilot plant stage, and several systems are being evaluated. If the development of solvent-refining technology proves technically and economically successful, it may obviate the need for flue gas desulfurization systems. However, solvent refining is likely to be a very expensive sulfur-removal technology.

Coal Gasification and Liquefaction

In addition to beneficiation and solvent refining, coal can be precleaned by conversion to gases or liquids. During coal conversion to synthetic fuels (synfuels), sulfur and ash are removed to produce a clean-burning fuel. These synfuel processes are, however, inherently less energy efficient than direct coal burning.

Coal gasification to date has been the subject of more research and development than coal liquefaction, and a number of government- or industry-sponsored coal gasification demonstration projects are under evaluation. The major obstacle to commercial use of synfuels is likely to be cost.

Although conversion to produce synthetic fuels looks promising for removing traditional pollutants such as sulfur and particulate matter, new health hazards may be created by large-scale coal-conversion facilities. These synfuel plants are likely to produce relatively high levels of polycyclic aromatic hydrocarbons, aromatic amines, toxic metals, and organometallic compounds.

Fluidized-Bed Combustion (FBC)

In conventional combustion systems, fuel is burned in a fixed bed or by suspension firing. In FBC, air is passed upward at high velocities from a distribution grid. The solids containing fuel, inert granular bed materials, and dolomite or limestone (sulfur sorbents) are held in turbulent suspension by the upward flow of air. This turbulent suspension appears diffuse, behaving like a liquid in that bubbles of gas rise through the bed as if it were boiling.

Combustion within the bed is intense. Coal burns so rapidly that the bed at any given time is composed almost entirely of the inert materials and sorbents present before combustion was initiated. Because of the intense combustion, high volumetric heat release is achieved. High heat-transfer rates are also obtained by the horizontal boiler tubes immersed in the fluidized bed. As a consequence, FBC can operate at lower temperatures. Optimum FBC temperatures are 750–950°C, whereas conventional utility boilers operate at temperatures as high as 1,400–1,500°C. These lower operating temperatures result in lower emissions of oxides of nitrogen.

Up to 90% of sulfur oxides can be removed by using limestone or dolomite as the bed material. The combustion heat converts limestone to lime, which reacts with SO_2, H_2O, and O_2 to produce calcium sulfite and sulfate. The reactivity of the sorbent can be maintained by the addition of new limestone.

Although FBC has been touted as a replacement for conventional boilers for nearly two decades, its use has been very limited. The transfer of this technology from the research and development stage to commercial use has been agonizingly slow.

Flue Gas Desulfurization (FGD)

Under NSPS (New Source Performance Standards) promulgated by the U.S. EPA in June 1979, FGD systems or scrubbers are required for all new or modified coal-fired utility boilers. For coal with sulfur contents in excess of 1.5%, 90% removal of SO_x is required; 70% SO_x removal is required for coals with sulfur contents of less than 1.5%. These percentage removal requirements reflect the economic and political desire to use high-sulfur, eastern U.S. coals. These standards also encourage the use of dry scrubbing technology (for low-sulfur coal) and coal beneficiation combined with FGD (for high-sulfur coal).

At the present time, FGD systems provide the most economically viable technology for achieving SO_x emission requirements. Since the 1970s, more than 50 FGD processes have been developed. Many have been tested in small-scale laboratory situations or in pilot plants.

There are two types of FGD systems: those that can be regenerated and those that are throwaway. In the first, the used-up, generally expensive sorbents are recovered by stripping SO_x from the scrubbing medium. Depending on the regeneration technology employed, useful products can be recovered in the form of elemental sulfur, sulfuric acid, liquefied SO_x, or gypsum. In throwaway systems, the scrubbing medium is inexpensive, and consequently recovery is not economically desirable. However, throwaway systems generally have significant waste-disposal problems.

FGD systems can also be classified as wet or dry. In wet processes, flue gases are scrubbed in a liquid or a liquid-solid slurry. In dry processes, solid sorbents are utilized. To date, wet scrubbing technology has received the greatest acceptance for FGD in utility boilers and other industrial applications. In any wet absorption processes, SO_x removal efficiencies of 90% or greater can be achieved. Presently, >90% of all utility FGD operating systems in the United States utilize a throwaway lime or limestone slurry process.

Lime or limestone slurry. In this FGD method, SO_x are absorbed in a slurry of lime or limestone. Slurry–flue gas contact can be achieved by several scrubber designs, including venturis, spray towers, marble beds, and turbulent contact systems. Major components in the FGD system include the scrubbing module and demister, a stack-gas reheater, and materials-handling and sludge-disposal units. Particulate matter may be collected simultaneously or removed by separate dry collection units upstream of the scrubber.

After SO_2 absorption, the scrubbing medium contains a mixture of calcium sulfite, calcium sulfate, and unreacted sorbent. This medium can be recirculated after some of the slurry is bled off and additional limestone is added. Waste slurry that has been removed is pumped to a settling pond for stabilization and ultimately disposed.

Although lime or limestone slurry systems can provide efficient SO_x removal (up to 98%), they have been plagued by a number of problems that have reduced their reliability and performance. One of these problems is the scaling and plugging of scrubber components by insoluble compounds. If deposits are excessive, they may interfere with the flow of gas or scrubbing medium, and the system may have to be shut down and cleaned. Another problem is the erosion and corrosion of metal surfaces by the scrubbing medium. Wet lime or limestone systems have high capital costs, and because of the energy-intensive nature of this technology, operating costs are also very high.

Dry scrubbing systems. Dry scrubbing represents a promising technology where 70% SO_x removal from flue gases is required from low-sulfur coals. Dry FGD can be achieved by bringing a dry alkaline sorbent such as lime, sodium carbonate, and naturally occurring carbonates into contact with flue gases. These sorbents can also be placed in solution or slurried and sprayed into flue gases, which vaporize the liquid. The mixture of fly ash, sulfates, and dry sorbent is collected and removed by dust collectors. These types of systems usually cannot be regenerated.

Dry systems have some important advantages over wet systems: 1) freedom from scaling problems and plugging; 2) no sludge-handling requirements; 3) less corrosion and therefore less maintenance; 4) energy requirements that are 25–50% lower; and 5) lower water requirements. Dry systems are advantageous in the low-sulfur coal-rich western states in the United States, where water resources are limited.

Air Pollution Control Through Government Regulation

The success of governmental regulatory programs depends on the selection of appropriate strategies. In the United States, four pollution control strategies or concepts are in use or have been proposed for the abatement of air pollution: emission standards, air quality standards, pollution taxes, and cost-benefit analyses. The first two provide the basic framework for most modern air pollution control programs. The others are more theoretical and have received relatively less attention.

Emission Standards

The basic purpose of emission standards is to establish limits on emissions for specific groups of sources, presuming that there is some maximum possible or practical degree of control. Once regulatory agencies determine the degree of reduction possible for a source group, then all members of that group may not emit more than the permitted level. If emission standards are interpreted to require the most practical degree of control, then cost will be an important consideration in attaining permitted emission limits and the best kind of technology or management practices to be utilized.

The "best practicable means" approach is commonly used for the development of emission standards. The term *practicable* implies economic and political as well as technological practicality. Using best practicable means may require the degree of control achieved by the best industrial plants in a source category or technology that can be reasonably applied from other industries. Implicit in this approach is that emission standards become more stringent as pollution control technology continues to improve. The best practicable means approach is the essence of emission limitations that require the use of "reasonably available control technology" (RACT). This is in contrast to the use of "best available control technology" (BACT), in which the emphasis is on reducing emissions with state-of-the-art technology without considering capital and operating costs.

Several other types of emission limitation are also in common use, including visible emission standards, prohibitive standards, limitations on fuel content, and numerical standards. Standards for visible emissions attempt to limit the **opacity** of stack plumes. Assuming that plume opacity and emission rate are related, emissions can be reduced by limiting the plume opacity. Prohibitive standards are designed to eliminate the practice of open burning. Fuel-content standards may be used to limit the amount of sulfur in oil and coal and lead or the reactive HC (hydrocarbon) content of gasoline. Numerical standards limit the level of emissions and may be based on a unit of time (kg/hr), air volume (g/dry standard m^3), heat input (kg/10^6 kJ), or weight of process material (kg/tonne).

Air Quality Standards Versus Critical Levels

In the efforts to assess the present and potential adverse impacts of air quality (e.g., O_3) on the environment (e.g., crops), scientists within the UN-ECE (United Nations Economic Commission for Europe) have applied the concept of critical levels. In contrast, in North America, ambient air quality objectives (Canada) or standards (United States) have been in use.

Critical levels are defined as concentrations of pollutants in the atmosphere above which direct adverse effects on receptors, such as plants, ecosystems, or materials, may occur according to present knowledge (UN-ECE, 1988; Bull, 1991; TemaNord, 1994). A critical level is thus based solely upon the best available scientific knowledge and understanding. In contrast, ambient air quality objectives or standards in North America, while based upon the best available scientific knowledge and understanding, are balanced against social, economic, and political considerations in the regulatory process. An additional difference between critical levels and ambient air quality objectives or standards is that while the critical levels approach explicitly assumes the existence of a "threshold" concentration above which there is a directly measurable adverse effect on sensitive components of the ecosystem, ambient air quality objectives or standards do not assume the existence of a threshold. Although one can debate the validity of utilizing the notion of a threshold concentration (Woodwell, 1975; Grigal, 1991), the idea of such a threshold is fundamental to the application of the concept of critical levels.

Another difference between critical levels and ambient air quality objectives or standards relates to how they are applied. Ambient air quality objectives or standards provide regulators with a measure of air quality for select air contaminants for compliance with legislation (International Union of Air Pollution Prevention Associations [IUAPPA], 1988). Critical levels are not used for compliance purposes. They are used as an integral part of pollutant emission control or abatement strategies. Initially, the geographic distribution of critical levels for selected sensitive ecosystem receptors are mapped. These data form the basis for testing emission control or abatement strategies with the objective of formulating an emission-control strategy that will reduce the ambient concentrations of the air contaminant of concern to a critical level or below.

Although one can see from the previous narrative that there are major differences between the UN-ECE's critical levels approach and the ambient air quality objectives or standards in North America, it is clear that the critical levels concept has merits for environmental protection. The most positive feature is that while the critical levels philosophy addresses scientific issues in a proactive fashion, it is not encumbered by social, economic, and political considerations. This is in strong contrast to the situation encountered in the process of evaluating, revising, and setting ambient air quality objectives or standards in North America, where social, economic, and political considerations have a major impact on the scientific agenda. Nevertheless, it should be recognized that a strong and unbiased scientific foundation is the only way to reduce uncertainty in ensuring protection of the environment from the adverse effects of air contaminants.

There is a long and somewhat convoluted history behind the critical levels approach to gaseous air contaminants, since it has evolved from the concept of critical and/or target loads relevant to acidic deposition. The critical levels approach is still evolving. It has been recognized, for example, that critical levels and critical loads for gaseous pollutants and acidic deposition, respectively, must be addressed together rather than separately. For additional details concerning critical levels, see UN-ECE (1988), Bull (1991), Grigal (1991), TemaNord (1994), Fuhrer and Achermann (1994), and Ashenden et al. (1995); and concerning ambient air quality objectives or standards, IUAPPA (1988).

Emission Taxes

In an emission tax strategy, emitters of major pollutants would be taxed according to a scale based on emission rate. The tax rate would be set so that emitters would find it more economical to control emissions than to pay the tax. Implicit in such an approach is that there would be no moral or legal sanction imposed against a source choosing to pay the tax rather than control its emissions.

Though the emission tax concept has been a point of discussion over the years, it has only recently been applied as a means of abatement of an air pollution problem. For example, regulations promulgated by the U.S. EPA to implement the goals of the Montreal Protocol include a substantial tax on the use of chlorofluorocarbons.

Cost-Benefit Analyses

In a cost-benefit approach, an attempt would be made to quantify pollutant-induced damages and the costs of controlling the pollutants responsible. It assumes that for damages, particularly health effects, there is no threshold value or that the threshold is so low that control at this level would be economically and politically impracticable.

In such analyses, it would be essential to accurately quantify both the cost of control and the cost of effects. Effects costs would be limited to tangible costs, including the dollar value of health care and mortality, crop losses, reduction in property values, and material losses such as cracking of rubber and paint and fading of fabrics. Intangible values such as the personal loss of friends and loved ones and loss of aesthetic values would be excluded.

In Figure 9.3, the costs of pollution control are compared with the costs of pollutant-induced effects and benefits of control. Note that as the cost of pollutant effects diminishes, the cost of control rises. Alternatively, as benefits rise, so of course do pollution control costs. Also note the steep rise in pollution control costs as effects costs decrease toward zero. This illustrates a major dilemma in pollution control. That is, effects costs decrease (or inversely, benefits increase) rapidly for a relatively small increase in control costs. As a greater degree of control is desired, however, control costs increase disproportionately to the benefits derived. In economic theory, the optimum degree of control would minimize effects costs (or inversely, maximize benefits) and also

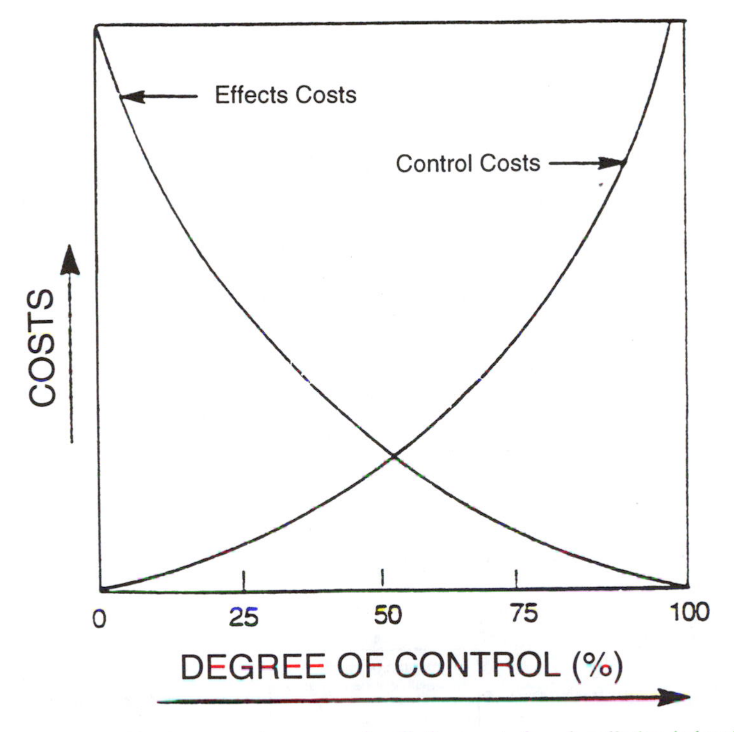

Fig. 9.3. Relationship between the costs of pollution control and pollution-induced effects. (Adapted from Godish, 1991)

minimize control costs. This would occur at the juncture of the two lines in Figure 9.3.

A Brief Assessment

In the final analysis, the air pollution system begins with the various sources of anthropogenic and natural emissions (Fig. 9.4). These are defined as primary pollutants since they are emitted directly into the air from their sources and include, for example, SO_2, NO_x, CO, Pb, organic

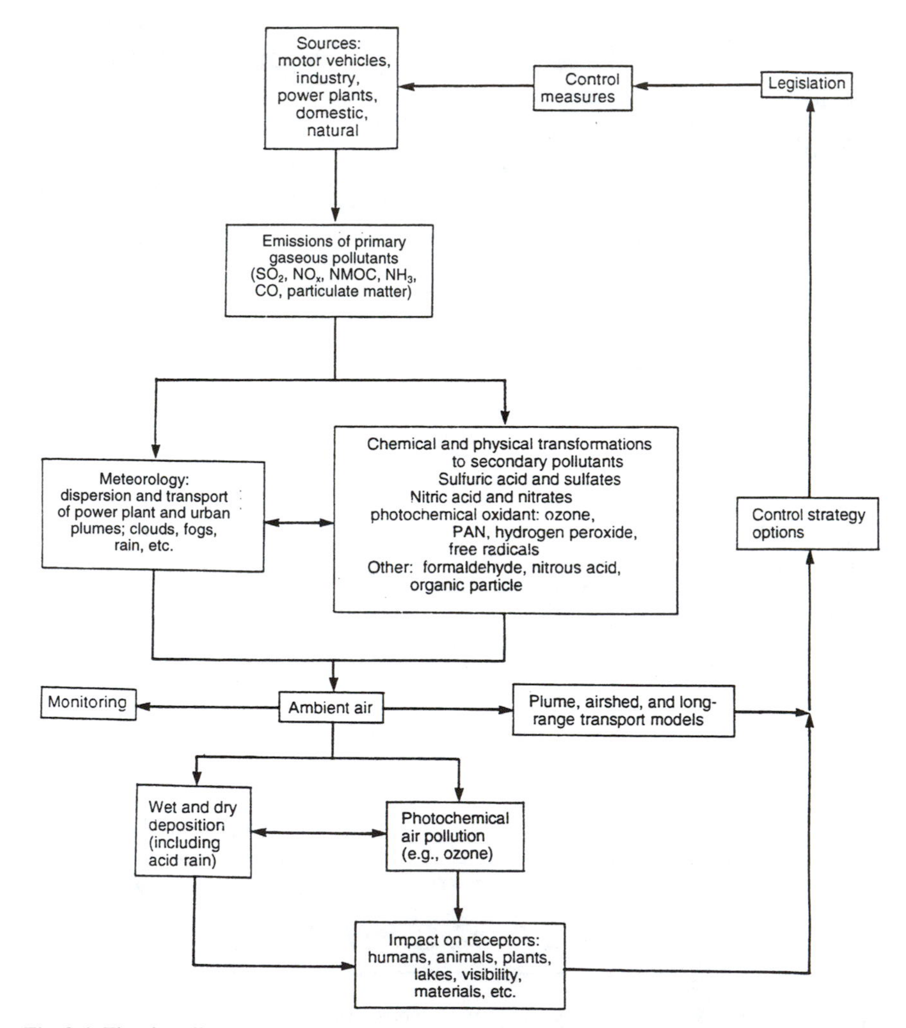

Fig. 9.4. The air pollution system. (Reprinted, by permission, from Finlayson-Pitts and Pitts, 1986)

compounds, and particulate matter. Once in the atmosphere, they are subjected to dispersion and transport (meteorology) and simultaneously to chemical and physical transformations to gaseous and particulate secondary pollutants, defined as those formed from reactions of the primary pollutants in air. The pollutants may be removed at the earth's surface via wet or dry deposition and can impact a variety of receptors, for example, humans, animals, aquatic ecosystems, vegetation, and materials.

From a detailed knowledge of the emissions, topography, meteorology, and chemistry, one can develop mathematical models that predict the concentrations of primary and secondary pollutants as a function of time at various locations in a given airshed. These computer-based models may describe the concentrations in a plume from a specific point source (plume models), in an air basin from a combination of diverse mobile and stationary sources (airshed models), or over a large geographic area downwind from a group of sources (long-range transport models). In order to validate these models, their predictions must be compared with the observed concentrations of the pollutants measured in appropriate ambient air monitoring programs; model inputs are then adjusted to obtain acceptable agreement between the observed and predicted values.

These validated models can then be used, in combination with the documented impacts on receptors, to develop various control strategy options. Finally, through legislative and administrative action, control measures can be formulated and implemented that directly affect the starting point of our air pollution system, that is, the primary emissions and their sources.

References

Ashenden, T. W., Bell, S. A., and Rafarel, C. R. 1995. Responses of white clover to gaseous pollutants and acid mists: Implications for setting critical levels and loads. New Phytol. 130:89-96.

Beaver, H. E. C. 1955. The growth of public opinion. Pages 1-11 in: Problems and Control of Air Pollution. F. S. Mallette, ed. Reinhold, New York.

Bull, K. R. 1991. The critical loads/levels approach to gaseous pollutant emission control. Environ. Pollut. 69:105-123.

Finlayson-Pitts, B. J., and Pitts, J. N., Jr. 1986. Atmospheric Chemistry: Fundamentals and Experimental Techniques. John Wiley & Sons, New York.

Fuhrer, J., and Achermann, B., eds. 1994. Critical Levels for Ozone: A UN-ECE Workshop Report. Swiss Federal Research Station for Agricultural Chemistry and Environmental Hygiene, Liebefeld-Bern, Switzerland.

Godish, T. 1991. Air Quality. 2nd ed. Lewis Publishers, Chelsea, MI.

Grigal, D. F. 1991. The Concept of Target and Critical Loads. Report to Electric Power Research Institute. Publ. EPRI EN-7318. Electric Power Research Institute, Palo Alto, CA.

International Union of Air Pollution Prevention Associations (IUAPPA). 1988. Clean Air Around the World: The Law and Practice of Air Pollution Control in 14 Countries in 5 Continents. I. Barker and J. Barker, eds. IUAPPA, Brighton, England.

TemaNord. 1994. Critical Levels for Tropospheric Ozone—Concepts and Criteria Tested for Nordic Conditions. Nordic Council of Ministers, Copenhagen.

United Nations Economic Commission for Europe (UN-ECE). 1988. ECE Critical Levels Workshop. UN-ECE Convention on Long-Range Transboundary Air Pollution.

Woodwell, G. M. 1975. The threshold problem in ecosystems. Pages 9-21 in: Ecosystem Analysis and Prediction. S. A. Levin, ed. Society for Industrial and Applied Mathematics, Philadelphia.

Further Reading

Hesketh, H. E. 1991. Air Pollution Control: Traditional and Hazardous Pollutants. Technomic, Lancaster, PA.

CHAPTER 10

Education, Research, and Technology Transfer: An International Perspective

Ignorance is a steep hill that perils rock at the bottom.

—Anonymous

Introduction

As sociopolitical and economic conflicts continue to escalate at the global scale, so too does the level of our awareness of the needs to reduce pollutant emissions into the atmosphere, conserve energy, and protect human health and welfare (the environment) against the adverse effects of air pollutants. However, the traditional view of addressing one environmental issue at a time (e.g., acidic rain or ozone) is largely intact. This compartmentalized view is primarily a product of our inability to address the integrative processes and products of the atmosphere as a whole and their impacts on life and material at the surface (Fig. 10.1). Clearly, this is a limitation and results from the complexity of the problem and the major financial support required to conduct the needed studies. A second limitation is the difficulty in assembling and developing cooperation among scientists representing many disciplines. Because of the competitive nature of securing funding for research, scientists in the United States and Europe continue to operate on a territorial basis, addressing single issues such as increases in air temperature and elevated levels of CO_2, O_3, or UV-B radiation. Environmental issues such as acidic precipitation and tree decline have been used to conduct numerous fragmented research studies ending in indecisive and/or less than dramatic results. With the rise of democratic governments in many previously communist countries and the present openness in those nations, evidence is starting to accumulate that human populations may have been exposed for years to relatively high atmospheric concentrations of various toxic chemicals such as complex organic pollutants and trace metals.

There are clear examples of the adverse effects of poor air quality on human health (e.g., carbon monoxide) and the environment (e.g., ozone). Conversely, at least for the moment, there is evidence that increases in the concentrations of certain atmospheric constituents such as CO_2 can benefit agronomic ecosystems when other growth-regulating factors (e.g., nutrients) are not limiting (Rogers et al., 1994). However, such elevated CO_2 levels can adversely impact fragile ecosystems such as the tussock tundra (Oechel et al., 1993). This brief discussion shows the need for a holistic understanding of the complex and dynamic inter-

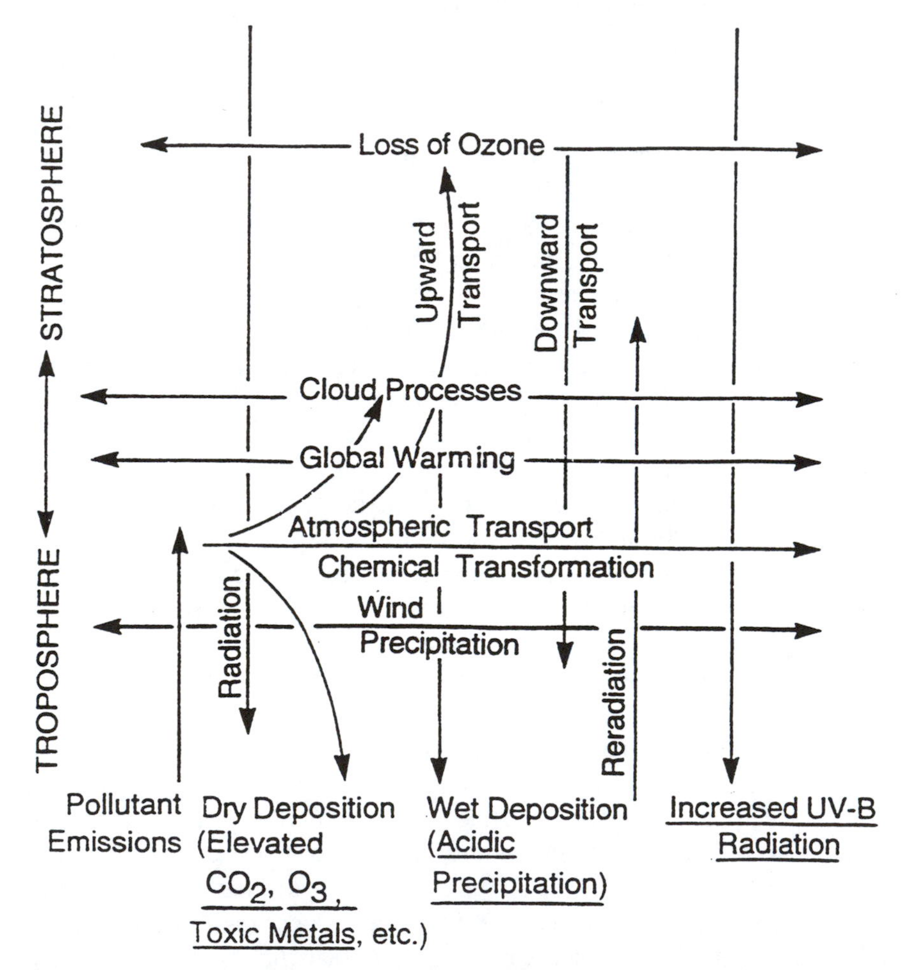

Fig. 10.1. Relationships between various physical and chemical processes and products in the atmosphere. (Reprinted, by permission, from Krupa and Legge, 1995)

actions between the multiple factors of the real world and the sensitive subjects responsive to poor air quality. It is important to realize that simply because many of us look for the simplest answers to complex real world problems, it does not necessarily mean that nature must cooperate. Therefore, as long as we approach integrated environmental problems with tunnel vision, we will continue to state the frequent conclusion that "more research is needed" to understand the problem and the corresponding solution. This highlights the conflict between those who believe in "wait and see, because we need more data" and those who believe in "why wait until a measurable human health or environmental impact occurs; control the potential cause now." Cost-benefit trade-offs represent a critical underpinning in this controversy. Three approaches to resolving air quality issues are mitigation, adaptation, and prevention.

Strategies for Improving Air Quality

Mitigation

Mitigation has been the most frequently used approach to improving air quality. In general, this approach is very costly, and thus cost-benefit trade-offs have played a major role. Air pollutant emission control technologies (see Chapter 9, Control Strategies for Air Pollution) and their application have essentially been the result of environmental laws or legislation, themselves a consequence of scientific or public pressure. At the present time, emission controls are in use mostly in developed countries. Developing nations have not readily embraced them because of the significant population growth in those countries, low standards of living, and economic pressure.

Although at the present time developed countries such as the United States are the largest per capita emitters of chemical constituents such as CO_2, future environmental laws may restrict such emissions through mitigation. As we move into the 21st century, developing nations may assume the role of major emitters of air pollutants. For example, the per capita emission of ozone in the Valley of Mexico has increased since the 1970s; but during the same period, ozone emissions in the Los Angeles area have declined (National Research Council, 1991; Miller, 1993).

Adaptation

Improvement of air quality through adaptation to modified lifestyles is in general an approach practical in developed nations. A prerequisite to the strategy of adaptation is **environmental literacy**, which is discussed in a later section of this chapter. While mitigation involves a specific source or sources, adaptation requires entire societies to change lifestyles. For example, the use of more energy-efficient lamps in all homes and businesses would reduce energy demand and thus lessen power production. Adaptation can be effective only if the required change is practiced across the board. The requirement by the U.S. EPA for the manufacture of and consequent societal adaptation to more energy-efficient automobiles is a clear example of the success of this approach. As with mitigation, because of economic pressures and environmental illiteracy, adaptation has not been successfully tested in developing countries. One example that may have a limited chance to succeed is the new practice of prohibiting a certain percentage of the population in Mexico City from driving their automobiles on certain days of the week to reduce mobile source emissions and thus the levels of photochemical smog. This requires adaptation by both the employer and the employee. In most urban centers in developed nations, significant progress has been made toward the development of mass transit systems. Such systems are prohibitively expensive for developing nations. Most importantly, a problematic approach is practiced in countries such as Mexico and India, where automobiles from the 1950s and the 1960s are still in operation because of continued repair of essential mechanical parts.

Prevention

Prevention is better than cure. Particularly in the United States, pollution prevention has been the theme of the 1990s. Pollution prevention requires changes in process technology. A simple example involves brick manufacturing. Conventional production of "bright red" brick can result in the emission of gaseous, toxic hydrogen fluoride gas. The manufacturing of "whitish pink" bricks, which contain high levels of calcium or an alkali, essentially absorbs the hydrogen fluoride. However, solving one type of problem can contribute to another environmental concern. Use of oxygenated fuels (e.g., ethanol mixed with gasoline) in automobiles leads to more complete fuel combustion and thus reduced emissions of carbon monoxide (which is toxic to humans and animals) but increased emissions of carbon dioxide (a greenhouse or global warming gas). Similarly, reducing sulfur dioxide emissions can lead to reductions in the extent of the acidic precipitation problem. Yet sulfur dioxide is a **cooling gas** in the context of global warming. These types of contradictory phenomena can cause difficulties in the strategies applied in pollution prevention. Yet significant and successful progress has been made in implementing pollution prevention, particularly in the chemical-manufacturing industry. Because of the types of complexities described as examples, pollution prevention requires not only significant advances in production technology, but also vast economic resources to make the needed changes. Again, these considerations limit its global-scale applicability at the present time.

Growing Concern about Poor Air Quality and Its Impacts on Human Health

A brief historical perspective of disastrous air quality episodes and their impacts on human mortality was provided in Chapter 3, An Historical Perspective of Air Quality. Such examples have been somewhat rare but not absent during recent times. Nevertheless, the chronic effects of poor air quality on human health continues to be a major issue as our knowledge of the subject grows. Urban pollution and its impacts on our health continue to be critical issues (National Research Council, 1991). Photochemical smog, toxic metals, and organic pollutants occupy a central theme. While Los Angeles smog prevails, urban pollution (including smog, particulate matter, and lead emissions from mobile sources) has reached critical levels at locations such as Manila in the Philippines. Approximately 30% of the citizens of Manila are known to suffer from bronchial problems and asthma and blood levels of lead that are disconcertingly high. The most recent evidence suggests that PM (particulate matter)-10 levels above 42 μg can be related to increased human mortality (R. K. Stevens, *personal communication*). Problems similar to those in Manila most likely occur in other urban centers in the developing nations but remain inadequately studied. The control of particulate matter from stationary sources, use of catalytic converters and unleaded gasoline in automobiles, and efforts to implement effective mass transit systems in the urban centers of developed nations have provided relief to those locations. Such strategies require stringent environmental laws, economic resources, environmental literacy, and societal adaptation. These represent critical limiting factors in their successful application in developing nations.

Global Climate Change Versus Global Change

Particularly since the mid-1980s, outstanding progress has been made in our understanding of the sources and sinks of chemical constituents in the atmosphere. Although many uncertainties remain, the traditional separatist views of physical and chemical climatology are rapidly merging (Fig. 10.1). Significant amounts of some atmospheric constituents such as methane come from natural sources, while others such as the chlorofluorocarbons are totally a consequence of human activities. While vegetation is a sink for atmospheric carbon dioxide during the day, it is a source during the night. All three of these gases are "greenhouse gases," and it is predicted that increases in their future concentrations will lead to increases in the global air temperature (i.e., global warming).

Although how much warming will occur and when and where it will occur are highly controversial subjects, increases in the concentrations of many of the atmospheric chemical constituents alone, their possible role in the destruction of the beneficial stratospheric ozone layer across all latitudes, and consequent potential increases in the deleterious ultraviolet-B radiation at the surface are all factors in "global climate change." Since human activity is known to be the major driving factor for the predicted global climate change, such a change will affect our lifestyles in the future and thus our impact on climate. There is a bidirectional feedback between the so-called global climate change and society (Fig. 10.2). Therefore, it is more precise to use the term **global change**

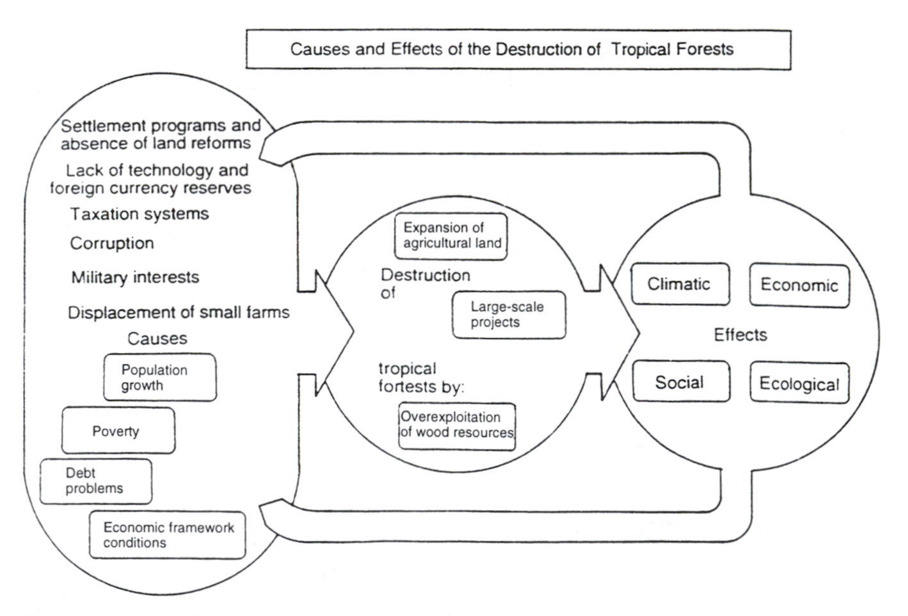

Fig. 10.2. Relationships between the causes and the effects of the destruction of tropical forests. (Reprinted from German Bundestag, 1991)

rather than global climate change. Reduction in world population is a critical component of global change, and world food supply and demand will be crucial in this context.

Balancing the World's Food Supply

At the present time in the United States, there is a surplus of food, although for economic and societal reasons, it is not distributed uniformly across all sectors of the population. Similarly, the former West Germany had the highest crop productivity per hectare in the world for many major crops because of very intensive management of the restricted land area under cultivation (Krupa and Kickert, 1993). On a nationwide basis in the newly unified Germany, this situation has changed somewhat because of the uncertainties associated with crop production as farming has changed from state-owned farms to private enterprise in the eastern sector. Land and property ownership in the present eastern sector of Germany is a controversial issue within the unified Germany. Land ownership prior to World War II or prior to the division of the former East and West Germany versus present ownership is the focus of much debate; many citizens who escaped from the east to the west can now return to the eastern sector and reclaim their previously owned property if they can provide legal documentation in support of proper ownership.

Although much progress has been made in agricultural production in developing countries, continued population growth, sociopolitical conflicts, and inefficient or corrupt distribution systems have contributed to a lack of uniform food supply across all sectors in those countries. Thus, starvation is rampant in some parts of the world, for example, in some African nations. This situation is expected to be affected further by global climate change. Continued increases in the atmospheric concentrations of carbon dioxide alone will require increased nutrient supply to sustain crop production. For example, under this scenario, phosphate fertilizer is already considered to be a limiting factor in some parts of Africa (Sombroek and Bazzaz, in press). Similarly, resource demand as a whole for crop production is expected to increase, and this again will most likely affect crop production in developing countries. Elevated atmospheric CO_2 levels coupled with any increases in air temperature and other growth-regulating factors is bound to alter the incidence of plant disease and insect herbivory. We have very little knowledge of these processes. If climatic changes occur slowly, plant breeders most likely can compensate for them. However, as accelerated plant breeding continues, genetic diversity relative to the wild type will be progressively compressed (Hoyt, 1988) to a point at which the magnitude of success may gradually decline. This simply means that certain crops grown in certain geographic areas may have to be replaced by others. For example, the corn grown in the corn belt of the United States could be replaced by grain sorghum, possibly because of increased air temperature and limited soil moisture.

The overall prediction is that developed nations will have to adjust and increase their food production in the future to compensate for any corresponding decline in the developing countries (Sombroek and Bazzaz, in press).

Preserving Biological Diversity

Although for decades ecologists have raised significant concerns about declining populations of certain flora and fauna, air quality and global climate change have provided another dimension to the issue. Dramatic shifts in biological diversity have frequently been a product of direct human intervention (e.g., the continued harvesting of the Amazon and Guatemalan rain forests). Excessive atmospheric inputs of nitrogen (mainly as ammonia or ammonium) have resulted in the invasion and overgrowth of the heather moors by tall grass in the Netherlands. Similarly, future increases in atmospheric CO_2 concentrations most likely could result in shifts in competition between C_3 and C_4 plants in mixed communities. Such shifts will alter the composition of native ecosystems and reduce the biological diversity of both producers and consumers. In essence, some species may disappear completely. Since there is feedback between various components in an ecosystem, loss of biodiversity will lead to altered ecosystems. The consequences of such changes are illustrated in Figure 10.2. The Endangered Species Act in the United States and similar laws in other nations protect flora and fauna against direct human abuse. However, such laws can sometimes lead to political conflict between nations when they involve cultural and economic questions or differences in philosophy relative to these issues.

Recently, two major national initiatives were established to assess and preserve biological diversity in the United States: 1) the Environmental Monitoring and Assessment Program (EMAP) and 2) the National Biodiversity Inventory (NBI). However, in the United States, these types of programs have not been sustained beyond a few years because of changes in governmental policies. The German parliament has undertaken a massive analysis of global climate change and potential impacts on global resources. The United Nations Food and Agricultural Organization (UN-FAO) is evaluating the impacts of global climate change on world agricultural production. These are some examples of short- and long-term assessments of the future biological diversity of both managed and native ecosystems.

Environmental Literacy

Environmental literacy requires a unique combination of knowing unbiased scientific facts and using them in a rational manner. Here, a little knowledge can be more dangerous than no knowledge. Although scientists contribute to the knowledge base in a technical sense, the media disseminate the information for public and political response. Because the public is rightfully concerned about human health and welfare, environmental issues, including air quality, frequently stimulate emotions that can be difficult to separate from scientific facts. The perception of risk and the actual risk can then be very difficult to separate. Even when the two phenomena are separated, the public can fail to accept the facts if emotions outweigh science. In some societies, industry-sponsored research may not receive public acceptance, even when it is correct, because of the historical distrust of such information. Traditionally, many profit-driven industries have sponsored defensive or reactive rather than proactive research, which is one of the reasons for public distrust of such research. One solution to this type of societal distrust is the model represented by the Alberta Government–Industry Acidic Deposition Research Program (Legge and Krupa, 1990) in which the provincial government and the concerned industry sponsored a large, integrated research program involving scientists from both public and private research sectors. The administrative aspects of the study were coordinated by a sponsors committee (government and industry), the actual science overseen by an independent science advisory board, and the results communicated by a nonpartisan, public advisory committee representing various interests. The end product of the research was peer reviewed and published in world literature (Legge and Krupa, 1990). Such efforts are rare, but they serve as models for resolving philosophical conflicts on sensitive environmental issues. These types of efforts can serve to improve our environmental literacy for problem resolution.

In developing nations, however, environmental literacy is directly correlated with lifestyles and a basic lack of education. Population growth, illiteracy, economics, lack of food, and the need for decent shelter outweigh environmental concerns (Fig. 10.2). In addition to the abuse of available natural resources, uncontrolled use of chemicals and poor industrial technology and operation are critical concerns. Although mitigation through political and economic pressure and technology transfer are possible in these cases, adaptation that requires environmental literacy is either virtually impossible or at best can be only partially achieved.

A completely different analysis is needed for those countries that have changed recently from socialist to democratic governments. The required science is available, at least in theory, but the pressures of economics and the adaptation to a market-driven economy are retarding factors. Another limitation is the lack of full knowledge of subtle but complex air quality issues (e.g., toxic chemicals), since air pollutant emissions have occurred in these countries unabated for decades and their ambient concentrations have not always been monitored in a scientifically defensible fashion.

Opportunities for International Cooperation

International cooperation does not necessarily mean sharing wealth, although some developing countries have tried to use this as a prerequisite for improving environmental conservation, as India did at the United Nations Conference on Environment and Development in Rio de Janeiro, Brazil, in 1992. While this view may have some validity, conservation of the global environment also requires the sharing of knowledge through education, technology transfer, and on-site remediation. It is important to note that, in general, many developing nations have competent scientists. These individuals simply need opportunities to apply their science and, more importantly, peers to communicate with on the scene. There is nothing better than local solutions to local problems, since they have a better chance of succeeding through local social acceptance.

There are a number of international agencies striving to deal with global-scale problems: 1) the World Bank; 2) the United Nations Environmental Program; 3) the UN-FAO; 4) the World Health Organization; and 5) various international institutions, such as the Commission of the European Economic Communities, the Rockefeller Foundation, and the U.S. Agency for International Development. There are many other similar organizations.

NATO (North Atlantic Treaty Organization), SCOPE (Scientific Committee on Problems of the Environment), and others have contributed significantly to the technical assessment of environmental problems. A disturbing fact, however, is the most recent shift in aid from developing countries to eastern-block countries that have converted from socialism to democracy. For example, there is some controversy about the U.S. administration's present emphasis on Poland, a country that has been more successful than some other eastern-block countries in shifting from socialism to democracy.

Although such help is needed, in the long run, this type of unbalanced support essentially defeats on-going programs. Such shifts are driven by short-term sociopolitical considerations. In the long term, the sharing of knowledge and education are sustainable commodities, and this is where academic professionals can play a critical role at the global scale. For example, the Peace Corps in the United States and similar programs elsewhere have contributed significantly to humanity across the world for decades, and the time has come for a similar program, the Peace Corps for Environmental Conservation. This would require the participation of environmental scientists from many nations and economic support from those nations.

At the present time, mitigation or on-site remediation is largely the domain of the private sector in developed nations. Such efforts need to be coupled with improved environmental literacy. The ideal approach to optimal success requires cooperation between the academic community and the private sector. There is much room for improvement. In some countries, such as Canada, government sponsorship of environmental research is greatly enhanced if academic communities can demonstrate cooperation with the private sector and potential economic or societal benefits to be obtained through technology transfer. We clearly need more cooperation of this type.

A Brief Assessment of the Limitations

Air quality and climate change issues are embedded in the conflict between environment and development. Many of the as-yet unresolved global problems, such as the population explosion, underdevelopment, poverty, and hunger, are currently escalating and are reflected in increasing environmental destruction.

About 80% of the world's energy-related emissions of radiatively active trace gases is currently caused by 15% of the world's population.

Energy consumption in the industrialized nations of the north has reached an all-time high. The per capita energy consumption in the developing countries is a fraction (between about 1/10 and 1/40) of that in industrialized nations (German Bundestag, 1991). It is foreseeable that as the developing countries follow the industrialization path of the developed nations, they will play a much greater role in the changes in our air quality and climate. This impact on the chemical and physical climate will be caused by more than just industrialization. The destruction of the environment in these countries (e.g., tropical deforestation and the conversion of deforested areas into farmland) results from poverty. Furthermore, since there are no other affordable fuels and no working energy-supply systems, forests are cut down in order to obtain firewood as a free and essential source of energy. The situation is dramatically aggravated by the population explosion currently observed in these countries. As a result, the overuse of environmental resources will increase.

Scientific and technological progress in the industrialized nations tends to accentuate economic differences between the rich and poor countries, and it tends to make it more difficult to introduce technological innovations into economically deprived nations. The position of developing countries in the world trade market is relatively weak. World market prices for their commodities are rather low. Their poverty level continues to increase because of high foreign debt, decreasing foreign investment within the developing countries, and a substantial net capital outflow from the poor to the rich countries. The gap between the north and the south is becoming wider (German Bundestag, 1991), and unless developing countries are given a fair chance to improve their economic status, it will be impossible to stop the destruction of natural resources, such as the tropical rain forests.

If air quality and climate are to be preserved, it will be necessary for industrialized nations to reduce their disproportionate polluting of the environment and for developing countries to overcome their socio-economic problems in an ecologically sustainable manner by achieving their own development in a manner consistent with their prevalent traditions and the conditions.

In their justifiable desire to satisfy the basic needs of their populations and to close the prosperity gap between themselves and the industrialized nations, the developing countries have so far been guided mainly by the economic systems of the industrialized nations, which have already led to the global overutilization of resources. Therefore, those making decisions about future international cooperation should consider these described limitations when designing environmental programs and coordinating scientific cooperation and technology transfer.

References

German Bundestag. 1991. Protecting the Earth. A Status Report with Recommendations for a New Energy Policy. Deutscher Bundestag, Referat Öffentlichkeitsarbeit, Bonn.

Hoyt, E. 1988. Conserving the Wild Relatives of Crops. International Board for Plant Genetic Resources, UN-FAO, Rome.

Krupa, S. V., and Kickert, R. N. 1993. The Effects of Elevated Ultraviolet (UV)-B Radiation on Agricultural Production. A critical assessment report submitted to the Formal Commission on "Protecting the Earth's Atmosphere" of the German Parliament. The German Parliament, Bonn.

Krupa, S. V., and Legge, A. H. 1995. Air quality and its possible impacts on terrestrial ecosystems in the Great Plains: An overview. Environ. Pollut. 88:1-11.

Legge, A. H., and Krupa, S. V., eds. 1990. Acidic Deposition: Sulphur and Nitrogen Oxides. Lewis Publishers, Chelsea, MI.

Miller, P. R. 1993. Response of forests to ozone in a changing atmospheric environment. Angew. Bot. 67:42-46.

National Research Council. 1991. Rethinking the Ozone Problem in Urban and Regional Air Pollution. National Academy Press, Washington, DC.

Oechel, W. C., Hastings, S. J., Vourlitis, G., Jenkins, M., Riechers, G., and Grulke, N. 1993. Recent change of Arctic tundra ecosystems from a net carbon dioxide sink to a source. Nature 361:520-523.

Rogers, H. H., Runion, G. B., and Krupa, S. V. 1994. Plant responses to atmospheric CO_2 enrichment, with emphasis on roots and the rhizosphere. Environ. Pollut. 83:155-189.

Sombroek, W. G., and Bazzaz, F. A., eds. Global climate change and agricultural production. In: Proc. UN-FAO Expert Consultancy. Wiley InterScience, London. (In press.)

Further Reading

Schneider, S. H. 1989. Global Warming: Are We Entering the Greenhouse Century? Sierra Club Books, San Francisco.

Trexler, M. C., and Haugen, C. 1995. Keeping It Green: Tropical Forestry Opportunities for Mitigating Climate Change. World Resources Institute, Washington, DC.

Glossary

< —less than
> —greater than
~ —approximately
cfm—cubic feet per minute
cm sec^{-1}—centimeters per second
dscf—dry standard cubic feet
grains/dscf—grains ($\times$ 0.0648 = 1 gram) per dry standard cubic feet ($ft^3 = 2.83 \times 10^{-2}\ m^3$)
kg—kilogram (= 10^3 gram or 2.2 pounds)
kg ha^{-1}—kilograms per hectare
km—kilometer (= 10^3 meter or 0.62 mile)
m—meter (= 1.094 yd)
mg/l—milligrams per liter
mg m^{-3}—milligrams per cubic meter
mm—millimeter
µg—microgram (= 10^{-6} gram)
µg g^{-1}—micrograms per gram
µg m^{-3}—micrograms per cubic meter
µm—micrometer (= 10^{-6} meter)
ng—nanogram (= 10^{-9} gram)
ng m^{-3}—nanograms per cubic meter
nm—nanometer (= 10^{-9} meter)
ppb—parts per billion (10^9) (= nl/l)
ppbC—parts per billion carbon equivalents
ppbv—parts per billion by volume
ppm—parts per million (10^6) (= µl/l)
ppmC—parts per million carbon equivalents
ppmv—parts per million by volume
ppt—parts per trillion (10^{12}) (= pl/l)

pptv—parts per trillion by volume
Tg—teragram (= 10^{12} gram)

absorption—removal of a substance (e.g., a gas) by passage through a liquid or liquid slurry
acid rain—generally considered as rain with a pH value of <5.68
acidic precipitation—rain, snow, sleet, etc. with a pH value of <5.68
acute exposure—exposure to relatively high concentrations of various air pollutants (e.g., SO_2, particulate matter, or methyl isothiocyanate) from a few hours to days to weeks
adsorbate—a gas or vapor contaminant that is trapped by an adsorbent
adsorbent—a solid medium capable of collecting gas or vapor contaminants
adventitious root—a root growing from an abnormal position on a plant, e.g., from the bottom portion of the stem or above the surface of the soil
aerosol—liquid or solid particle in a gaseous (air) atmosphere
air pollutant—a chemical constituent added to the atmosphere through human activities resulting in the elevation of its concentration above a background level
air pollution—contamination of the atmosphere, especially with human-made waste (e.g., an air pollutant)
air quality guideline—an air quality regulation for a pollutant set by a government to protect human health and welfare
albedo—the fraction of incident electromagnetic radiation reflected by a surface
Ambient Air Quality Standard—*see* air quality guideline
angina pectoris—a disease marked by brief, sudden attacks of chest pain in humans precipitated by deficient oxygenation of the heart muscle
angiosarcoma—a rare form of liver cancer in workers occupationally exposed to polyvinyl chloride (PVC) plastics
anthropogenic activity—human activity
area source—a large urban center, e.g., Minneapolis-St. Paul, MN
asbestosis—industrially contracted lung disease in humans characterized by a diffuse fibrosis or scarring in the lower lobes of the lungs
asthma—a condition in humans often marked by labored breathing accompanied by wheezing
ataxia—inability of humans and animals to coordinate voluntary muscular movements
atmospheric inversion—temperature increase with altitude at some height above the surface (aloft)

banding—bands of yellow tissue alternating with green tissue on conifer needles or on grass leaves
berylliosis—multiple nodulelike accumulation of inflamed tissue in the lung in which the nodules exhibit granulation; enlargement of the lymph nodes in humans
bifacial—pertaining to both sides of a leaf
biological indicator—sensitive plant species such as tobacco or bean, which shows typical foliar injury symptoms and can be used as an indicator of relative air pollution
bronchitis—inflammation of the bronchial tubes in humans
bronzing—bronze coloring on the upper (e.g., caused by O_3) or lower (e.g., caused by PAN) surface of a leaf
Brownian motion—the random motion of microscopic particles suspended in a liquid or gas caused by collision with molecules of the surrounding medium
Btu—British thermal unit; the quantity of heat required to raise the temperature of 454 g (1 pound) of water at or near its point of maximum density (3.9°C, 39.1°F) 1°F; 1 Btu = 0.252 kilocalorie

C_3 plant—a plant of almost all crops of temperate origin in which the assimilated CO_2 is fixed in a molecule composed of three carbon atoms
C_4 plant—a plant in a crop of tropical origin, such as sorghum or sugarcane, in which the assimilated CO_2 is fixed in a molecule composed of four carbon atoms
carboxyhemoglobin (COHb)—carbon monoxide (CO) bound with blood hemoglobin (Hb)
CFC—chlorofluorocarbon
chlorosis—loss of chlorophyll or yellowing of a leaf
chronic exposure—exposure to low-level concentrations of various air pollutants with intermittent, periodic, random episodes of high concentrations that last from a few minutes to one or more days
cloud seeding—release of AgI (silver iodide) particles into the air to cause cloud formation and generate rain
coarse particle—particle >2.5 μm in diameter
cold front—a line of intersection on the earth's surface at which cold air overtakes and flows beneath warm air
condensation nucleus—a particle in the atmosphere on which moisture can condense, e.g., to form clouds
continental source—many adjacent and closely located nations, each with many urban centers, e.g., Europe

continuous point source—a chimney at a given geographic location that emits air pollutants continuously, e.g., a coal-fired power plant

criteria pollutant—a pollutant that has documented effects on people, plants, or materials at concentrations, or approaching those, found in polluted air

diffusion—the gradual mixing of the molecules of two or more substances as a result of random thermal motion

downwind—the direction toward which the wind is blowing; with the wind

dry deposition—deposition of air pollutants onto surfaces during periods of no precipitation

edema—an abnormal, excess accumulation of watery fluids in the connective tissue of humans

emphysema—a condition in humans marked by enlargement of the lung and frequently by impairment of heart action

epidemiology—the study of the sum of factors controlling the presence or absence of a particular disease

epinasty—a physiological process by which a plant part (e.g., a petiole) is moved outward and downward like a shepherd's crook

episode—a period (minutes, hours, days, or weeks) of relatively high air pollutant concentrations

eq—equivalence; molecular weight ÷ valence of an ion (allows the calculation of an ion balance of the amount of anions [–] against cations [+] in a solution, e.g., rain)

erythema—a redness of Caucasian skin, as that caused by sunburn

feeder root—a secondary, fine root on a tree that is essential for water and nutrient absorption. These roots are not woody and, unlike the primary root, are not involved in anchoring the tree.

$FEV_{1.0}$—forced expiratory (release of air from human lungs) volume in the first second of a vital capacity maneuver

filtered air—ambient air filtered through a particulate filter and an activated charcoal filter to remove dust and O_3 in air pollution–plant response studies

fine particle—particle <2.5 μm in diameter

fleck—a spot or mark, usually white, gray, or brown, on the upper leaf surface in which the tissue becomes dry or elevated

fluorosis—a disease caused in animals (rarely in humans) by fluoride accumulation

friction—resistance to relative motion between two bodies in contact, e.g., two molecules in the atmosphere

FVC—forced vital capacity, i.e., the amount of air that can be forcibly expelled from human lungs after a full inspiration

general circulation—the movement of jet streams toward the poles from west to east north of the equator and from east to west south of the equator caused by the Coriolis force or the influence of the earth's rotation

general circulation model (GCM)—a computer climatological model based on general circulation patterns of the upper air

gradient winds—the winds associated with isobars (regions of equal atmospheric pressure)

gravitational settling—the falling of matter onto surfaces as influenced by gravity

greenhouse effect—a term commonly used to describe the heating of the earth's surface air through the blockage of outgoing radiation by trace gases in the atmosphere, acting like the roof on a greenhouse

hν—light energy

heterogeneous reaction—a reaction in the atmosphere from gas to particle phase (particles may be dry or wet, e.g., cloud water or fog droplets)

high pressure system—an atmospheric system with barometric pressure >760 mm of mercury

homogeneous reaction—a reaction in the atmosphere from gas to gas phase

inhalable particles—particles <10 μm in diameter that can enter and be deposited in the respiratory systems of humans and animals

interveinal—between veins on a leaf

isobars—geographic regions of equal high or low atmospheric pressure

isothermal—pertaining to constant temperature with change in height above the surface

LEL—lower explosion limit, i.e., the level below which an open flare will not stay in flame

line source—an air pollutant source from which pollutant emission occurs in a line, e.g., a highway with automobiles traveling in a line or an aircraft jet stream

long-range transport—transport of air pollutants from hundreds to thousands of kilometers downwind from a source area

low pressure system—an atmospheric system with barometric pressure <760 mm of mercury

lysosome—a saclike cell organelle that contains various hydrolytic enzymes (enzymes that split a bond and add the element of water)

M—matter, a third body in a reaction that involves two primary substances and that regulates the energy in the reaction

m MSL—meters above mean sea level

macrophage phagocytic activity—activity of a large cell that engulfs foreign matter in humans and animals

mesopause—the transition zone between the mesosphere and the thermosphere in the earth's atmosphere

mesophyll—a layer of cells on the upper leaf surface with chlorophyll-containing chloroplasts that is responsible for photosynthesis or carbon dioxide assimilation (also known as the palisade layer)

mesosphere—area 50 to ~85 km above the earth's surface

mesothelioma—a tumor derived from the mesothelial tissue such as the membrane folded over the surface of the lung in humans

mobile source—a source of air pollution that is capable of motion, e.g., a car

monocot—a plant with a single cotyledon, e.g., a grass species

mottling—blotches of yellow color (chlorosis) mixed with the normal green surface of a conifer needle; loss of tooth enamel in humans or animals

multipotent carcinogen—a chemical that produces cancers in many different organs and in some cases produces different kinds of tumors in the same organ in humans and animals

necrosis—death of leaf tissue, which usually turns brown, dark brown, red, or black

nonfiltered air—ambient air filtered for dust only in air pollution–plant response studies

normal adiabatic lapse rate—temperature decrease at a rate of 1°C/100 m (5.4°F/1,000 ft) above the earth's surface

novel forest damage—recent (e.g., 1970s–1980s) phenomenon of extensive forest damage (also called *neuartige Waldschäden*)

occluded front—point at which cold and warm fronts merge (low pressure will be associated with the merger)

olefin—straight-chained or branched hydrocarbon (made of hydrogen and carbon) containing one or more double bonds

organic pollutant—a pollutant containing carbon atoms, e.g., polyaromatic hydrocarbons (PAHs)

oxygenated fuel—fuel rich in oxygen used for combustion in motor vehicles

ozone hole—a hole caused by the loss of O_3 from the beneficial O_3 layer in the stratosphere

peroxyacetyl nitrate (PAN)—an organic pollutant formed in the atmosphere by photochemical processes

photochemical oxidant—a substance produced through photochemical or light-driven chemical reactions in the atmosphere and capable of oxidizing another that is present in a reduced form

photochemical processes—light-driven reactions, e.g., those involving O_3 (ozone) and molecular O_2 in the atmosphere

photochemical reaction—a chemical reaction driven by sunlight

photochemical smog—a combination of smoke and fog, produced mainly through photochemical reactions

photodissociate—to dissociate in the presence of light

photolytic—pertaining to chemical decomposition in the presence of light

phytotoxic—toxic to plants

PM-10—particles <10 μm in diameter

polar vortex—an atmospheric vortex at either of the poles caused by the difference in the direction of the circulation of the jet streams north and south of the equator

polyaromatic hydrocarbon (PAH)—a compound that is derived from a six-carbon benzene ring containing several rings of carbon and hydrogen atoms in its structure (several members of this group are known carcinogens)

primary particulate matter—particulate matter emitted by a source

primary pollutant—a pollutant emitted directly into the atmosphere by a source, e.g., sulfur dioxide, SO_2, or nitrogen dioxide, NO_2

pulmonary—pertaining to the lungs

pulmonary compliance—ability of the lung to yield to increases in pressure without disruption

regional source—several closely located urban centers, e.g., Los Angeles and the entire South Coast Air Basin in California

respirable—particles <2.5 μm in diameter that can enter and be deposited in the pulmonary system

Salmonella—a bacterium used in testing for human and animal carcinogenicity of organic pollutants

secondary particulate matter—particles formed in the atmosphere from primary pollutants, e.g., SO_4^{2-} particles

secondary pollutants—pollutants formed secondarily in the atmosphere through reactions involving primary emissions, e.g., ozone (O_3) and sulfate (SO_4^{2-}) particles

silvering—a process whereby leaf color turns from green to silver

simulated acidic rain (SAR)—an artificial mixture of chemicals used to mimic ambient acidic rain in experiments

single-event point source—an accidental chemical spill at one time at a given geographic location, e.g., derailment of a chemical container

smoke damage—the classic type of forest decline observed in the past in Germany (also called *Rauchschaden*)

spirometry—the measurement of the volume of air entering and leaving the lungs

stagnant air mass—an air mass that is stagnant because of the presence of a stationary front

stationary front—the combination of regional-scale cold and warm air masses

stationary source—a pollution source that remains at one place, e.g., a coal-fired power plant

steady state reaction—a reaction in which the concentrations of the substrates and the products remain in equilibrium

stipple—a condition in which groups of cells about the size of a pin or nail head accumulate brown to dark red pigment at several locations on a leaf surface

stomata—openings on the lower surface of foliage to facilitate gas exchange and transpiration

stork's nest—a shape similar to that of a nest built by a stork on top of a tree in which to lay its eggs

stratopause—the transition zone between the stratosphere and the mesosphere in the earth's atmosphere

stratosphere—the area approximately 15–50 km above the earth's surface

subadiabatic lapse rate—temperature decrease at a rate of $<1°C/100$ m ($<5.4°F/1,000$ ft) above the earth's surface

subsidence—inversion aloft that facilitates a descending motion of air in the atmosphere and usually implying that the condition extends over a rather broad area

sulfurous smog—a combination of smoke and fog composed mainly of SO_2, particulate matter, and moisture

superadiabatic lapse rate—temperature decrease at a rate >1°C/100 m (>5.4°F/1,000 ft) above the earth's surface

surface-based inversion—temperature increase with increase in height from the earth's surface

suture—a seamlike joint on a fruit, e.g., a peach

thermosphere—the area approximately >85 km above the earth's surface

tidal volume—the rise and fall in amount of air intake and changes in respiratory mechanics in humans and animals

tropopause—the transition zone between the troposphere and the stratosphere in the earth's atmosphere

troposphere—the area up to about 10–12 km above the earth's surface

TSP—total suspended particulate matter in the atmosphere

UEL—upper explosion limit, the level above which an open flare will not stay in flame

upwind—in or toward the direction from which the wind is blowing

UV—ultraviolet radiation

UV-A—UV band at 320–400 nm

UV-B—UV band at 280–320 nm

UV-C—UV band at <280 nm

warm front—line of intersection on the earth's surface at which warm air overtakes and flows above cold air

wet deposition—deposition of air pollutants onto surfaces by precipitation (rain, fog, hail, etc.)

Index